BTEC
Entry 3/Level 1

D0352669

CONSTRUCTION

Simon Topliss | John Murray-Smith

Student Book

ENTRY LEVEL
3/1

A PEARSON COMPANY

Published by Pearson Education Limited, a company incorporated in England and Wales, having its registered office at Edinburgh Gate, Harlow, Essex, CM20 2JE. Registered company number: 872828

www.pearsonschoolsandfecolleges.co.uk

Edexcel is a registered trademark of Edexcel Limited

Text © Pearson Education Limited 2010

First published 2010

13 12 11 10
10 9 8 7 6 5 4 3 2 1

British Library Cataloguing in Publication Data
A catalogue record for this book is available from the British Library.

ISBN 978 1 84690 920 7

Edited by Bruce Nicholson
Designed by Pearson Education Limited
Typeset by Tek-Art
Index by Indexing Specialists (UK) Ltd.
Cover design by Pearson Education Limited
Cover photo © Corbis: Blend Images/Klaus Tiedge
Back cover photos © Shutterstock: rtem (right), Stanislav Duda (left)
Printed in Italy by Rotolito Lombarda

Disclaimer
This material has been published on behalf of Edexcel and offers high-quality support for the delivery of Edexcel qualifications.

This does not mean that the material is essential to achieve any Edexcel qualification, nor does it mean that this is the only suitable material available to support any Edexcel qualification. Edexcel material will not be used verbatim in setting any Edexcel examination or assessment. Any resource lists produced by Edexcel shall include this and other appropriate resources.

Copies of official specifications for all Edexcel qualifications may be found on the Edexcel website: www.edexcel.com

Contents

About the authors

Simon Topliss is a lecturer and team leader in construction at the Grimsby Institute of Further and Higher Education. He has taught BTEC Nationals and BTEC Higher Nationals in Construction and helped develop many of the new QCF units with Edexcel. He has previously written books that support the BTEC First, BTEC National and Diploma construction qualifications. He is a trainer for Edexcel and a Senior Standards Verifier.

John Murray-Smith is currently working as curriculum manager for Schools Construction at Havering College of Further and Higher Education. He has previously worked at a number of other colleges in London and the south east of England. He has considerable construction industrial experience and has worked as far afield as west Africa.

Credits

The publisher would like to thank the following for their kind permission to reproduce their photographs:

(Key: b-bottom; c-centre; l-left; r-right; t-top)

Alamy Images: Art Director and Tripp 104c, Catchlight Visual Services 99, David J Green 55t, 115b, Eye-Candy Images 93, Fancy 150, Image Source 30, 51, 82, Mar Photographics 17, Metalpix 137/5, Mike Goldwater 35t, Moodboard 28, Paul Drake 113b, Paul Dronsfield 104b, Richard Heyes 136/3, Studiomode 101/4, 112/4, UK Stock Images Ltd 133; **Construction Photography**: Jean-Francois Cordella 9, Phil Starling 36; **Corbis**: Jose Fuste Raga 1; **Getty Images**: PhotoDisc 65tr, 77, 124/3; **iStockphoto**: Stefan Klein 108; **London 2012**: 4; **Masterfile UK Ltd**: Jerzyworks 68; **Pearson Education Ltd**: Ben Nicholson 126b, Clark Wiseman, Studio 8 20, 115t, David Sanderson 18/1, 18/2, 34, Gareth Boden 16c, 16b, 27, 40, 52/1, 52/2, 52/3, 52/4, 52/5, 62/1, 62/2, 62/3, 62/5, 62/6, 63/1, 63/2, 63/3, 63/5, 64l, 64r, 65tl, 67, 72/2, 72/4, 72/5, 73/1, 73/2, 73/4, 73/5, 74, 86t, 86c, 86b, 87/1, 87/2, 87/3, 87/4, 100/1, 100/2, 100/3, 100/4, 100/6, 101/1, 101/2, 101/3, 101/5, 103, 111, 112/5, 112/6, 113c, 114, 117, 124/1, 124/2, 124/5, 125t, 136/1, 136/4, 136/5, 136/6, 136/7, 137/4, HL Studio 47, Trevor Clifford 16t, 62/4, 63/4, 72/6, 73/3, 112/1, 112/3, 126t; **Photolibrary.com**: Imagebroker 151; **Robert Down Photography**: 55b; **Science Photo Library Ltd**: Maxine Adcock 135; **Shutterstock**: 6493866629 123, 125b, auremar 39, 48, Barry Barnes 18/3, Brian Weed 75, Chiran Vlad 85, Christina Richards 31, Creation 19tl, Daleen Loest 80, Dancing Fish 18/4, Dwight Smith 11, Elnur 54b, 146, gabor2100 19br, hannahmariah 96, Igor Pshenin 124/6, Igor Shikov 19bl, Ilya Andriyanov 12, Joe Belanger 72/3, Keith WIlson 112/2, Lack-o-Kean 19tr, Monika Wisniewska 145, Mountain Light Studios 136/2, Nata-Lia 124/4, Natalia V. Guseva 54t, nikkytok 137/3, Norman Pogson 100/5, Petrov Stanislav 88, Photoseeker 142, rtem 61, Scouting Stock 35b, Sean Prior 7, Stanislav Duda 90, Stefana Tiraboschi 125c, Steve Carroll 72/7, tinamo 113t, Tokio Maple 72/1, Vladislav Gajic 71, wacpan 137/2, Wheatley 137/1, Zodchiy 65b; **Travis Perkins plc**: 89

Cover images: *Front*: **Corbis**: Blend Images/Klaus Tiedge; *Back*: **Shutterstock**: rtem tr, Stanislav Duda tl

All other images © Pearson Education

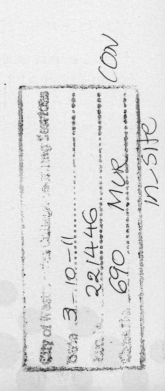

About your BTEC Entry 3/Level 1 Construction

Choosing to study for a BTEC Entry 3/Level 1 Construction qualification is a great decision to make for lots of reasons.

The UK construction industry employs over 2 million people with thousands more in many linked industries. It is a career that can take you around the world working on many interesting and exciting construction projects.

A career in construction enables you to leave your own mark on the built environment, as you would be working on projects that will be there for years to come. Their construction would be helped through your contribution.

Your BTEC Entry 3/Level 1 Construction is a vocational or work-related qualification. It will give you the chance to gain knowledge, understanding and skills that are important in the subject or area of work you have chosen.

What will you be doing?

This book covers enough units for you to gain any of the following qualifications:

- BTEC Entry 3 **Award** in Construction

- BTEC Level 1 **Award** in Construction

- BTEC Level 1 **Certificate** in Construction

- BTEC Level 1 **Diploma** in Construction

If you are unsure, your tutor will let you know what level of qualification you are aiming for.

How to use this book

This book is designed to help you through your BTEC Entry 3/Level 1 Construction course. It is divided into 12 units to match the units in the specification. Each unit is broken down into smaller topics.

As you can see on the next pages, the book contains many features that will help you get the most from your course.

Introduction

Each chapter starts with a page that gives you a snapshot of what you will be learning from that unit.

STARTING WORK IN CONSTRUCTION
UNIT 9

Construction is an exciting industry to work in and has many career opportunities ranging from managers to the bricklayers on site. Construction work is undertaken all over the world, from major bridge projects and dam building, to the world's tallest sky-scraper in Dubai. This type of work requires many different roles to be filled with trained people. A career in construction is an exciting opportunity for a young person to enter and one in which you can work anywhere in the world.

In this unit you will learn:

• About the different types of career opportunities available in construction

• About the different types of organisations offering career opportunities in construction

• How career choices can impact upon your lifestyle

• How to make informed career choices

• To work responsibly with others

• To ask for and respond to guidance when working as part of a team.

This is the world's tallest building, the Burj Dubai. Would you want to work at a height like this?

Activities

You will find activities throughout the book. These will help you understand the information in the units and give you a chance to try things for yourself.

Activity: Career choice

There are many websites that provide details of the different types of construction jobs. Find one website and have a look through the job descriptions, making notes of which ones you are interested in. Look back at the activity *My ideal career* and see if the ideas you listed will be provided by the jobs you are interested in.

Case studies

Case studies show you how what you are learning applies in the real world of work.

Case study:
Nick, project manager

This is Nick. He is a project manager with a large construction company. He looks after projects that run into millions of pounds of investment. Nick has made his informed career choice by trying different contractors on a self-employed basis, before working directly for this company.

• **Would you like Nick's job?**

• **Would you be prepared to travel around the UK?**

• **How would you balance your work with a family life?**

Functional skills

Useful pointers showing you where you can improve your skills in English, mathematics and ICT.

Functional skills

While you improve your communication skills you will also develop your skills in **English**.

Key terms

The words you need to understand are easy to spot, and their meanings are clearly explained.

✳ Key term

Responsibly
Sensible and with regard for others.

Remember!

Look out for these boxes. They point out really important information.

❗ Remember

People often get stressed at critical times. Good communication will help these situations.

Check

You'll find a reminder of key information at the end of each topic.

✓ Check

- There are a variety of ways you can be employed in construction
- Professional and technical personnel have specific roles
- Skilled roles usually start with an apprenticeship

Assessment page

This page will help you check what you have done so far and give you tips for getting the best grade you can for each task.

Assessment overview

This table shows you what assessment criteria you need to meet to pass the unit and on which pages you will find activities and information to help you prepare for your assignments.

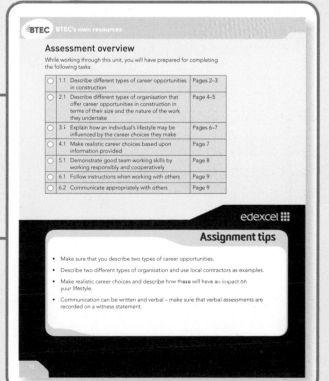

Edexcel's assignment tips

At the end of each chapter, you'll find hints and tips that will help you get the best mark you can.

Your book is just part of the exciting resources from Edexcel to help you succeed in your BTEC course. Visit www.edexcel.com/BTEC or www.pearsonfe.co.uk/BTEC2010 for more details.

STARTING WORK IN CONSTRUCTION

Construction is an exciting industry to work in and has many career opportunities ranging from managers to the bricklayers on site. Construction work is undertaken all over the world, from major bridge projects and dam building, to the world's tallest sky-scraper in Dubai. This type of work requires many different roles to be filled with trained people. A career in construction is an exciting opportunity for a young person to enter and one in which you can work anywhere in the world.

In this unit you will learn:

- About the different types of career opportunities available in construction

- About the different types of organisations offering career opportunities in construction

- How career choices can impact upon your lifestyle

- How to make informed career choices

- To work responsibly with others

- To ask for and respond to guidance when working as part of a team.

This is the world's tallest building, the Burj Dubai. Would you want to work at a height like this?

L01 Construction career opportunities

Types of career

There are many different ways of working within the construction industry, namely:

- Full-time
- Part-time
- Employed
- Self-employed
- Permanent
- Temporary

There are many professional and technical jobs in construction. Here is a list of some of the common ones:

- Architect
- Architectural technician
- Construction manager
- Maintenance manager
- Facilities manager
- Civil engineer
- Surveyor
- Structural engineer
- Building services engineer

The skilled craft roles which you enter by starting an apprenticeship in construction are:

- Bricklayer
- Carpenter
- Joiner
- Painter and decorator

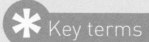

Key terms

Self-employed
Working for yourself, rather than being employed by a company.

Skilled
A trade skill that has taken a period of training to achieve.

- Plumber

- Electrician

- Plasterer

- Roofer

The majority of the manual unskilled work on site is undertaken by general operatives.

Activity: Careers within construction

The following are professional and technical careers within construction. Link each one to the correct box that identifies what it does.

Architectural technology	Heating
Surveying	Land measurement
Civil engineering	Load calculations
Structural engineering	Design
Building services engineering	Road construction

Check

- There are a variety of ways you can be employed in construction

- Professional and technical personnel have specific roles

- Skilled roles usually start with an apprenticeship

L02 Construction organisations

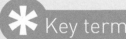 **Key term**

Contractor
A company that has been given a contract to do part or all of a construction project.

SMEs
Small and medium enterprises.

This is the London Olympic venue that is currently under construction. It is such a big project that more than one large **contractor** has to undertake the work in order to complete it on time.

Types of companies

There are many different types of construction companies operating in the UK.

These include:

- Large contractors

- **SMEs** – small and medium enterprises.

The larger construction companies will take on the multi-million value contracts and the smaller companies will take on local construction works. Often, on major works such as the Channel Tunnel, two or more larger contractors will work together as one contractor on its own cannot manage the whole contract.

Activity: Research

Have a look around your local built environment and find a large construction project. Find out who is undertaking the work and look at their website to research about them and the type of work that they do. Then answer these questions:

- Do they work nationally?

- What type of work do they produce?

Work undertaken

The range of construction work undertaken is vast, as the spider diagram below shows.

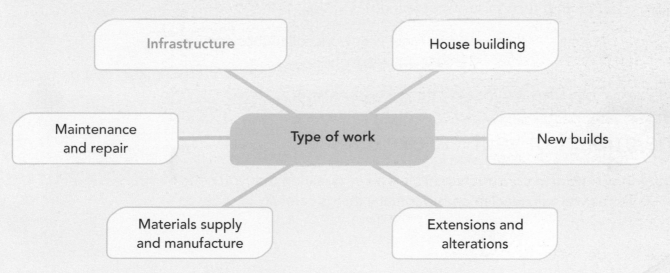

Infrastructure

House building

Maintenance and repair

Type of work

New builds

Materials supply and manufacture

Extensions and alterations

Activity: Types of infrastructure work

Think of three types of work that would come under infrastructure, for example motorways.

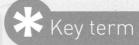
✓ Check

- There are different sized companies that undertake construction work

- The range of work undertaken in construction is extensive

L03 L04 Career choice & lifestyle

Lifestyle issues

The career that you want to follow in the future can have an effect on your lifestyle. Financial rewards, by earning money, are not the only thing that you will want to work for. Other things that will influence your career choice will be:

- *Ambitions and aspirations* – What do you really want to achieve in your life?

- *Job satisfaction* – Will you want to leave a successful project for a client with a sense of pride?

- *Sense of identity* – Will you want to belong to a professional organisation that recognises what you do?

- *Social benefits* – Will you want to meet with work colleagues outside of work?

- *Financial benefits* – What salary do you want to work towards?

The impact of your career choice

What job do I want in construction? This is a key question that you need to plan for, as you will need to consider many things carefully in order to come to a decision.

These include:

- Education
- Training
- Physical health
- Working conditions
- Flexible working hours
- Long hours
- Working in another country

The impact of these career demands can have an effect on several different aspects of your life, including:

- Your family
- Your social life
- Your personal relationships
- The amount of stress that you will be subjected to
- The mental and physical demands on your body.

Activity: My ideal career

Pick up a plain piece of paper and, leaving the centre blank, write around it ideas of what you want to do with your life in a construction career. What type of lifestyle do you want? What working conditions would you accept?

Brainstorm answers to these questions then select the career that meets these answers by placing its name in the centre of the page.

Case study:

Nick, project manager

This is Nick. He is a project manager with a large construction company. He looks after projects that run into millions of pounds of investment. Nick has made his informed career choice by trying different contractors on a self-employed basis, before working directly for this company.

- **Would you like Nick's job?**
- **Would you be prepared to travel around the UK?**
- **How would you balance your work with a family life?**

Activity: Career choice

There are many websites that provide details of the different types of construction jobs. Find one website and have a look through the job descriptions, making notes of which ones you are interested in. Look back at the activity *My ideal career* and see if the ideas you listed will be provided by the jobs you are interested in.

✔ Check

- The lifestyle you want depends on many different influences
- Your future career choice may involve many different work demands
- Having as much information as possible makes a career choice easier

L05 L06 Working responsibly with others

Behaviour

Construction sites can be very dangerous places to work. There have been many fatal and major accidents in construction, because, for example, temporary structures such as scaffolding are used where falls from height can occur.

Construction workers need to act **responsibly** and safely with regard to themselves and other people they work with. When working on a construction site you need to know:

- What you are responsible for
- Who your team members are
- What cooperation is needed in a task.

Key term

Responsibly
Sensible and with regard for others.

Activity: Strengths and weaknesses

Take a good look at yourself and under two column headings write down your *strengths* and *weaknesses* with regard to acting responsibly. The table below gives one example of a strength and one example of a weakness to start you off.

Strengths	Weaknesses
I am good at listening to what other people say	I don't always stand back and assess things before starting a job

Remember

- **Yes, you can enjoy a construction job, but you must act and think responsibly**
- **It is the simplest of mistakes that can cause an accident**

Working within a team

Working as part of a team is an essential part of delivering the project to a client. Large projects can involve a great number of people on site, as well as other teams within the **sub-contractors** that will be undertaking some of the work. Working in a team is rewarding as you will develop life-long friends, earn respect and enjoy working.

To become a good team member you will need to demonstrate things such as:

- *Enthusiasm* – How keen are you in working?

- *Approachability* – Can anyone talk to you easily?

- *Communication* – Can you receive and understand instructions?

Key term

Sub-contractors
These are contractors who work for the main builder and provide specialist services such as electrical installation works.

Activity: On-site communication

Communication is essential on a construction site. Instructions are an important part of that communication. Name one instruction that might be used on site for each of these types of communication:

- Visual

- Spoken

- Hearing.

Functional skills

While you improve your communication skills you will also develop your skills in **English**.

Check

- Working within a team involves a number of different skills

Assessment overview

While working through this unit, you will have prepared for completing the following tasks:

○	1.1	Describe different types of career opportunities in construction	Pages 2–3
○	2.1	Describe different types of organisation that offer career opportunities in construction in terms of their size and the nature of the work they undertake	Page 4–5
○	3.1	Explain how an individual's lifestyle may be influenced by the career choices they make	Pages 6–7
○	4.1	Make realistic career choices based upon information provided	Page 7
○	5.1	Demonstrate good team working skills by working responsibly and cooperatively	Page 8
○	6.1	Follow instructions when working with others	Page 9
○	6.2	Communicate appropriately with others	Page 9

Assignment tips

- Make sure that you describe two types of career opportunities.

- Describe two different types of organisation and use local contractors as examples.

- Make realistic career choices and describe how these will have an impact on your lifestyle.

- Communication can be written and verbal – make sure that verbal assessments are recorded on a witness statement.

HEALTH & SAFETY & WELFARE IN CONSTRUCTION

Health and safety and welfare on a construction site are very important and must involve everyone. Construction work can be very dangerous if it is not properly controlled by all involved. There are many hazards on construction sites that can cause fatal, major or minor accidents. Many temporary structures are used until a construction project is finished, all of which contain hazards that must be reduced to prevent accidents. In the UK there are many laws that concern health and safety that are used to protect all workers.

In this unit you will learn:

- About the causes of accidents in construction

- About the hazards and safety signs on a construction site

- How to minimise the risk of hazards

- The different types of fire extinguishers and when to use them

- About the legislation HASAWA and COSHH.

When might a construction worker have to wear a safety harness?

L01 Why do accidents happen?

Causes of accidents

There are many different causes of accidents in construction, many of which can be made by the simplest of mistakes. These can include:

- Falling – from working at height or into an excavation

- Electricity – electric shocks

- Manual handling – lifting and moving awkwardly

- Poor maintenance – failure of a piece of equipment

- Misuse – forcing something or not using the correct equipment to protect yourself

- Poor behaviour – messing about

- Housekeeping – an untidy site with trip hazards

- Confined spaces – a small space with no air

- Drugs and drink – causing you to be too confident

- Tiredness – falling asleep near a hazard

- Weather – sunburn from working unprotected outside or slipping on ice.

It is unfortunate that a few construction accidents do result in the death of the worker involved. However, all accidents can be prevented if the right precautions are taken. For example, don't rush, wear the right Personal Protection Equipment (PPE) and ensure the safety of yourself and others at all times.

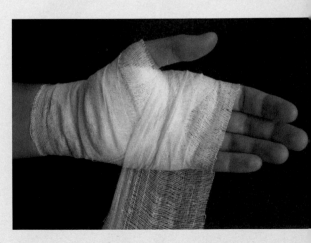

The worker in the photo (right) has had an accident. The accident means that work will stop on site while an investigation is carried out into how it happened and how to prevent it from happening again.

Activity: Accident investigation report

This is a report from an accident investigation but some words are missing. Complete the report by selecting the three correct words from the list at the end.

Work had started on the walls of the building by the bricklayers. They were working on a _____ that was supporting them and the bricks while they laid them along the wall. Someone had _____ a section of handrail and as a bricklayer stepped back to inspect their work, he _____ through the gap to the ground. Fortunately the bricklayer fell onto soft ground and was off work for only three days with a back injury.

platform	cut
removed	fell
scaffolding	

Activity: Hot weather on site

Think about how you could keep safe on site during hot weather.

Check

- There are many different causes of accidents in construction
- Many accidents can cause fatal or major injuries to workers

L02 Construction hazards 1: Materials

Hazards in construction can be anywhere – from the tools and equipment that you use, the ladders that you stand on and the materials used to produce the construction project or item. Hazards cannot be simply left – they must be dealt with and controlled so they cannot cause an accident.

We will examine some of the hazards that you will possibly meet while making your construction timber product in the workshop environment.

Materials

There are many different materials used in construction, all of which have very different hazards connected with them.

Material	Hazard
Sawn timber	Splinters and dust created when timber is sawn which can cause breathing problems
Fixings	Nails normally have to be driven in using a hammer and your hands and fingers could be injured doing this
Medium Density Fibreboard (MDF)	MDF has a glue in it which, when you cut it, can cause breathing problems as dust enters the air
Lime	Lime can cause burning and irritation to the eyes
Paints	Some paints contain solvents which can cause an irritation if inhaled or swallowed
Solders	These are used in plumbing operations and can burn the skin if still hot
Lead	Lead is present in old paint, flashings and old water pipes. If absorbed into the body, it can cause damage to the brain
Asbestos	This material is banned from use. It can cause lung and breathing problems as fine asbestos fibres become stuck in the lung tissues.

✳ Key terms

Environment
The surrounding area you are working in.

MDF
Medium Density Fibreboard.

Activity: Hazards in the workshop

Look around your workshop and identify a possible hazard from each of the following:

- Timber

- Machinery

- The environment.

Check

- Hazards in construction can be anywhere

- You need to be aware of potential hazards in the materials you work with

L02 Construction hazards 2: Tools & equipment

There are a vast number of tools used in construction, from woodworking and bricklaying through to plumbing and electrical. We will examine a few of the hand tools that you would use for these types of craft trades.

Tools

Tenon saw

This is used to cut fine joints in timber products. It has sharp teeth that cut across the timber grain and which can cut the skin if they come into contact with it.

Wood chisel

Various sized chisels are used to cut out timber from joints or to form recesses for fixings such as locks. They are extremely sharp and will badly cut your skin if your hand slips while using one.

Bolster

A bolster is used to cut bricks. It has a guard on the top.

Hammer

A bolster has to be used with a hammer, which can also be used to knock nails and other items into timber.

Pipe welding

Pipe welding is soldering using a blow torch. This melts the solder so that it forms a joint.

Equipment

The following types of equipment are used to help workers gain height so they can reach a work area. Each of them brings hazards.

Step ladders

These are available in various sizes and have a series of steps.

Hop-ups

These are small platforms used at low level that allow you to hop onto them to gain height.

Trestles

Trestles are used along with scaffold boards to provide a working base which can be adjusted for height. Additional handrails have to be fitted.

Activity: Platform hazards

Workers use platforms to gain extra height when they cannot reach high enough. Take a look at the workers on this platform. What are the hazards?

Activity: Access platform investigation

The use of the three methods above and opposite to gain height all need the use of handrails to hold on to while working and to help prevent workers from falling off the platform. Use the internet to find an access platform supplier and find out how handrails can be added to each of the three platforms (step ladders, hop-ups and trestles).

Functional skills

Researching for information on the internet will help you develop your skills in **ICT**.

✔ Check

- There are many different methods of gaining height. Each must be assessed for risks

There are many different safety signs used in construction to warn workers about hazards and dangers. The type of sign is indicated by its colour. There are some examples on the next two pages.

L02 Safety signs

Key terms

Mandatory
Must be followed.

Prohibited
Not allowed.

Blue and white

This is a mandatory sign

Red and white

This behaviour is prohibited

Yellow and black

This is a warning sign

Green and white

This is a safe condition sign

Activity: Name the sign and its meaning

Name the signs below and describe what they mean.

Check

- There are many hazards on construction sites
- The colour of a safety sign has a meaning

LO3 Controlling hazards

Risk assessment

A risk assessment is a document that lists all the main hazards and identifies which are the most dangerous. It then uses a control to lower the risk down to an acceptable level.

Current UK safety laws expect a risk assessment to be carried out on any commercial construction work. This is done to ensure that:

- The hazard is identified
- The risk is acceptable
- The controls are correct for the risk
- The risk assessment is reviewed.

The risk assessment is then communicated to the workers who must follow its instructions to reduce the effect of any hazards.

Personal Protective Equipment

Personal Protective Equipment (PPE) should always be the last thing that is considered in reducing the harm from hazards, as you need first to try and reduce the risk down to an acceptable level before using PPE. PPE covers things such as:

- Safety boots
- Gloves
- Hard hats
- Safety glasses.

This worker is wearing all the correct PPE for the cutting task on the brick.

Key term

PPE
Personal Protective Equipment.

Work planning

Any task on a construction site must be well planned to ensure that:

- The correct materials are used

- The work is done safely

- The risks are minimised.

Safe use of stepladders

Ladders should always be used correctly by ensuring that:

- They are on a level surface

- The ladders are at the correct height

- They are locked in the open position

- There is no damage to the ladders.

Activity: Risk assessment

Using the very simple headings on the form below, identify three hazards in your practical tasks. Think about what risk could result from the hazard and what you need to do to reduce or control the hazard.

Hazard	Risk	Control

Check

- Risks should be controlled so they are reduced

- There are a range of actions that can be taken to control hazards

L04 Fire extinguishers

Fires at work are a hazard that can prove fatal due to the toxic smoke that may be given off. Fire extinguishers are a quick method of fighting the fire once it has started and to extinguish it before it takes hold. There are different types of fire extinguisher that can be used.

Key term

Extinguish
To put out a fire or flame.

Types and uses of fire extinguishers

Fires need three things to start – air, fuel and heat (or an ignition source). This is called the triangle of fire.

The type of fire extinguisher that you use to put out the fire will depend on the type of fuel that the fire is feeding upon. When a construction site has been built there is a lot of packaging and waste materials present. If these are not cleared away they can become fuel for a potential fire. Sources of the start of the fire can be:

• Plumbing work with a blow torch

• Electrical fault

• Temporary heaters

• Vandalism.

The common types of fire extinguisher and their uses are shown in the following simple table.

Type of fire	Fire extinguisher (colour)	Fire extinguisher (type)
Paper and wood	Red band	Water
Electrical and liquids	Blue band	Powder
Liquid (not electrical)	Cream band	Foam
Liquid and electrical	Black band	Carbon dioxide

! Remember

- You cannot operate a fire extinguisher safely unless you have been trained to do so.

Activity: Fire!

A fire has broken out on the construction site. It was started by an electrical fault on a 110 volt transformer which has melted and is now setting fire to the floor it was stood upon.

- Are there any precautions you should take before starting to put out the fire?

- What type of fire extinguisher will you use to fight the fire?

- What is its colour coding?

✔ Check

- There are lots of different types of fire extinguishers

- You cannot operate a fire extinguisher safely unless you have been trained to do so

L05 Safety legislation

The law is used to protect workers from inadequate safety provisions by an employer. Safety legislation sets down minimum standards that an employer must meet.

The Health and Safety at Work Act 1974

The Health and Safety at Work Act 1974 (HASAWA) requires that employers:

- Provide employees with PPE

- Make sure that plant and equipment is safe

- Inform and supervise work

- Keep the workplace safe and healthy.

The Health and Safety at Work Act also contains duties for employees. The Act states that the employee has to 'take reasonable care for the health and safety of himself and of other persons who may be affected by his acts or omissions at work'.

✻ Key terms

HASAWA
Health and Safety at Work Act 1974.

COSHH
The Control of Substances Hazardous to Health regulations.

Substances
Chemicals, liquids and any material that could harm.

Activity: HASAWA

What do you think is meant by the sentence 'take reasonable care for the health and safety of himself and of other persons who may be affected by his acts or omissions at work'?

The Control of Substances Hazardous to Health regulations 2007

The Control of Substances Hazardous to Health regulations (COSHH) deal with many of the chemicals and substances that we use.
For example:

- Glues and adhesives

- Paints

- Dry materials.

Under COSHH employers have to assess the risks in using a product and train and inform employees and workers in its safe use. Manufacturers supply safety data sheets on all of their materials.

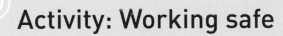

Activity: Working safe

PVA glue is used in the workshop for many timber jointing tasks. Find a container of glue and read the instructions on its side. Alternatively, find the supplier's website and download the safety data sheet. When you have done this, answer the following questions:

- Can the glue be used without wearing gloves?

- Is it harmful if swallowed?

- What precautions are needed if you get it on your skin?

Check

- Safety laws are there to protect the workers

- The use of chemicals and substances must be risk assessed

Functional skills

Reading instructions will help you develop your skills in **English**.

Assessment overview

While working through this unit, you will have prepared for completing the following tasks:

○	1.1	Describe the causes of accidents in construction	Pages 12–13
○	2.1	Identify potential hazards in a construction environment	Pages 14–17
○	2.2	Identify the safety signs used in a construction environment	Pages 18–19
○	3.1	Identify and select methods used to minimise the risks of hazards	Pages 20–21
○	4.1	List different types of fire extinguishers	Pages 22–23
○	4.2	Describe when different types of fire extinguishers should be used	Pages 22–23
○	5.1	Describe the purpose of HASAWA	Page 24
○	5.2	Describe the purpose of COSHH	Page 25

edexcel

Assignment tips

- Look at the website of the Health and Safety Executive which has lots of information on the causes of accidents.

- Look around your workshop – there should be lots of safety signs.

- Examine the methods used in your workshop to control hazards.

- Identify the different fire extinguishers around your building and their type.

- Look at a material safety sheet to help you describe the COSHH requirements.

WORKING AS A TEAM TO MOVE & HANDLE RESOURCES

A construction project uses many different materials to produce the final building. All of these are different shapes and sizes. Some can be moved by hand, but others may require large cranes to move them around the site. The movement of materials using hands and arms is known as manual handling and it involves the methods that you use to pick up and move an object. This has to be done properly so you do not injure yourself in the process.

In this unit you will learn:

- About the regulations and guidance that apply to the safe moving and handling of resources

- How unsafe manual handling techniques can cause injury to self and others

- To work as part of a team when carrying out safe moving and handling of resources

- To work responsibly with others

- To ask for and respond to guidance when working as part of a team.

Is this worker using a correct lifting method?

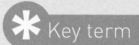
L01 Regulations & guidance

Manual handling regulations

The regulations on moving and handling were brought in to try and stop the number of days lost to injuries, mainly to the back, due to incorrect lifting and moving. The cause of over a third of accidents that involve three days, or more, off work is down to bad **manual handling**.

Days lost through injury cost the employer and the government in sickness benefits and claims, so any law that can prevent this is a good idea.

The regulations cover:

- Avoiding any unsafe manual handling

- Assessing handling risks

- Reducing the risks from handling.

Injury can have a very bad effect on a worker's colleagues, as well as their family, who might be affected by:

- Disability

- Long-term sickness

- Money worries.

Safe systems of work

Managers put into workplaces a **SSW (safe system of work)**. This means that careful thought has been put into place to avoid accidents to workers. Typical SSWs on manual handling would be:

- No movement of materials unless by forklift

- Gloves to be worn at all times during movement

- No worker to lift more than 20kgs.

A SSW must always be looked at to see that it is working correctly, that it reduces the risks from manual handling and is reviewed to make sure that any changes are working.

Object weights

You will always need to find out how much an object weighs. This is because if you start to lift it and it is heavier than you thought, you may cause yourself a serious injury. Also, the weight of an object will need to be known if it is going to be lifted by mechanical means because:

- It may exceed the safe working loads of the machine

- A specialist company may be required

- A means of holding the object may need designing, for example for sheets of glass.

Activity: Safe manual handling

The questions below on manual handling would be asked by a health and safety advisor to test your knowledge. Try and identify an answer before looking them up.

- What is the safest recommended load you can carry at arms length?

- What year were the manual handling regulations brought into force?

- What people do the regulations cover?

Check

- Manual handling regulations were brought in to reduce injuries through lifting

- Any lifting needs risk assessing

L02 Manual handling

Equipment

Workers must make proper use of equipment provided for their own safety when manual handling. Typical equipment provided for manual handling of construction materials would be:

- Sack barrows

- Wheel barrows

- Pallet trucks

- Forklift trucks.

Cooperation

Cooperation is essential when manual handling materials because often you may need more than one person – or a 'team lift' approach may be required. This could be due to:

- The weight of the material

- The size of the material

- The distance it has to be moved.

During your practical units you must show that you can cooperate with your tutor's instructions on health and safety matters.

Risks

Any manual handling must be looked at to ensure that it does not cause any injury to:

- The person carrying out the movements

- Any other person in the area

- Work colleagues.

The only method of doing this is by undertaking a risk assessment. Refer to *Unit 10 Health and safety and welfare in construction* which shows you how a simple risk assessment should be undertaken.

✱ Key term

Sack barrows
A two-wheeled trolley with two handles.

Pallet
The timber base that is used to lift and move items by forklift.

! Remember

Tell your tutor if you see any bad manual handling practices, so no injury occurs.

Case study:

Working safely

Paul had worked in the joinery shop for over ten years as a machinist on joinery items. The joinery shop had a materials store on the ground floor. Plywood materials were stored here and moved using a forklift truck up to the cutting table. A rush job had come into the workshop and the forklift truck was busy, so Paul fetched the sheet materials and carried them to the work station. Later on that day Paul complained of a sore back and pain and was sent home.

- Why did the accident happen?

- How could it have been prevented?

✓ Check

- Injury due to poor manual handling is one of the largest causes of time off work

- Before an item is moved it must be assessed for the possible risk

L03 Team manual handling

Safe moving

Any material, equipment or plant that you move must be assessed for its weight.

The guidelines from the **HSE** (Health and Safety Executive) are that any weight:

- Should be lifted close in to the body

- Should weigh no more than 25kg for a male and 16kg for a female.

When using any equipment to lift you should always read the instructions as to how much it will safely lift. When lifting in pairs it is not the case of half the weight of the object each, as one end may be heavier than the other, or you may be taller than the other person so it tips their way.

Key term

Team
Working with more than one person.

HSE
Health and Safety Executive.

! Remember

These are guidelines only – each individual is built differently and will be able to lift more or less safely.

Weight

The maximum amount you can lift will depend on a number of things (see below).

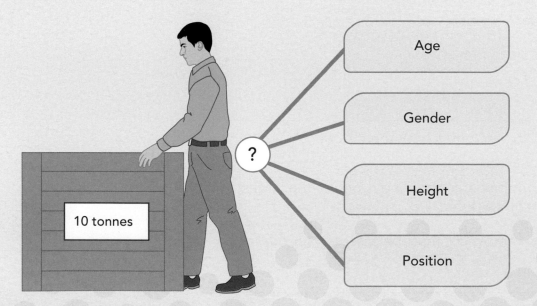

10 tonnes

? — Age
Gender
Height
Position

Height is not just the total height of your body, it can also be:

- The position of your knees

- The height of your elbows

- The height of your shoulders.

Activity: Team lift

You have been asked to lift a joinery item which weights 259 kg using a **team** of four people. How much will each person lift?

PPE

It may be necessary to use PPE (Personal Protection Equipment) to protect yourself when carrying out manual handling duties.

Activity: PPE identification

Identify the PPE required for each of these manual handling operations:

- Lifting concrete paving slabs

- Carrying bags of cement

- Wheeling mortar in a barrow.

Check

- Each person can lift a different amount depending on ability

- The weight of an object must be assessed

L03 Lifting

The photograph shows the correct way to lift an object.

Follow these steps when lifting:

- Set feet apart

- Place leading leg forward

- Bend the knees

- Grip object firmly

- Keep back straight

- Keep load close into body

- Keep shoulders level.

This lifting can be done in pairs, but a guide is shown in the following figure.

✳ Key term

Forklift trucks
A mechanical means of lifting using two forks.

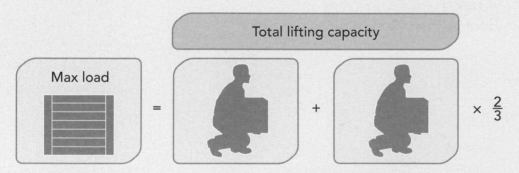

Maximum load = the number of people lifting × two thirds.

Lifting aids

There are many things that can help you lift heavy objects:

- Shelf trolleys – a wheeled trolley with shelves in it

- Sheet trolleys – used to carry sheet materials

- Sack barrows – used to carry materials in sacks or bagged

- Forklift trucks – can lift any material on a pallet

- Pallet trucks – hand-operated pallet lifters

- Sheet hoists – used to lift large sheets mechanically

- Chutes – used to move materials downwards

- Roll cages – a cage to hold materials that can be wheeled

- Lifting hooks – a hook built into the material for mechanical lifting.

Activity: Lifting aids

The following are used to lift various items. Can you name what they would lift?

Check

- There are a number of factors that must be considered before making a lift

- A number of aids are available to lift items

L04 L05 Working responsibly with others

Behaviour

When you work with other people in a team you have a 'duty of care' to take responsibility for both yourself and the team's safety. This is a legal requirement under the HASAWA (see *Unit 10 Health and safety and welfare in construction*). When manual handling you should demonstrate good behaviour by:

* Looking for any hazards and reporting them

* Cooperating with other team members

* Good clear communication

* Understanding and following instructions.

Not behaving properly when carrying out heavy lifting of awkwardly shaped materials could result in an accident to yourself or your fellow workers.

✻ Key term

Behaviour
The way in which you act – your conduct or manner.

Communication
Talking and listening.

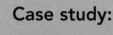

Case study:

Planning a team lift

A team lift was planned to move a manufactured joinery item out of the workshop onto a lorry and into a new office building. A team was put together to lift the joinery item. Four people were going to lift the joinery item and the manager had talked through the lift in detail. The weight was assessed and all four workers started to lift and move the item out of the workshop. As they were halfway to the lorry, one person dropped the load onto another worker's foot, causing a bad sprain above the ankle protector on their boots.

* What went wrong with this team lift?

! Remember

People often get stressed at critical times. Good communication will help these situations.

Respond to guidance

To work responsibly within a team, you need to show that you are:

- Enthusiastic – that you are keen and willing to take part

- Approachable – anyone can talk to you openly.

You also need to be a good communicator, which will involve:

- Taking time to listen to what someone else has to say

- Being able to ask questions and be understood clearly

- Being able to follow written and verbal instructions.

When you are manual handling in a team the instructions often have to be shouted. You will need the right attitude to react quickly and to respond efficiently.

Activity: Non-verbal communication

In teams of three, practise communicating the instruction 'The first aid box is in the office' to the other two people in your team without using your voice or writing the instruction.

Functional skills

Communicating by talking will help you with your functional skills in **English**.

Check

- Good communication is needed in team manual handling

- Good behaviour is needed to prevent accidents

- Good communication is needed when working in teams

Assessment overview

While working through this unit, you will have prepared for completing the following tasks:

○	1.1	Identify the regulations and guidance that apply to the safe moving and handling of resources, including the Manual Handling Operations Regulations 1992	Pages 28–29
○	2.1	Explain how unsafe manual handling techniques can cause injuries to themselves and others	Pages 30–31
○	3.1	Perform safe moving and handling of resources as part of a team, without the use of lifting aids	Pages 32–33
○	3.2	Perform safe moving and handling as part of a team, with the use of lifting aids	Pages 34–35
○	4.1	Demonstrate team working skills by working responsibly and cooperatively when moving and handling resources with others	Page 36
○	5.1	Follow instructions when working with others	Page 37
○	5.2	Communicate appropriately with others	Page 37

edexcel

Assignment tips

- When you practise lifting, use a light object.
- Look at the HSE website for guidance on manual handling.
- Assess the risks in lifting.
- Find out how far the object has to be moved.
- Ensure good communication when team lifting.

DEVELOPING CONSTRUCTION DRAWING SKILLS

The design of a building has to be produced by the skills of a designer using drawing techniques. The designer takes an idea and develops it into a 2D and 3D drawing that everyone can understand.

Basic drawings are produced using drawing equipment, a board, paper and pencil. Detailed drawings can also be produced on a computer monitor using a mouse or light pen.

In this unit you will learn:

● About the basic equipment used to produce construction drawings

● To prepare a sheet of drawing paper

● To produce a basic construction drawing

● To work responsibly with others

● To ask for and respond to guidance when working as part of a team.

Why are drawing skills important in many construction jobs?

L01 Construction drawing equipment

This is a basic drawing tool kit that is required to produce drawings.

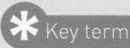

The drawing board

You need a plain flat surface to draw upon. In the photograph is a 't' shaped piece of equipment which is known as a 't-square'. This is used to run up and down the edge of the board. It enables a line to be drawn that is at 90 degrees to the vertical and so is perfectly horizontal. Any **set square** that is then placed onto the t-square edge will produce a line that is at 90 degrees to this.

This arrangement allows you to draw straight and accurate lines to form the outlines of buildings which have mostly 90 degree corners. The two set squares have different angles in them. They are 45/90/45 and 30/60/90 degrees.

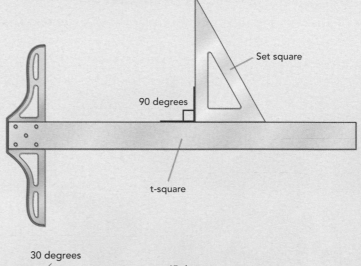

Set square

90 degrees

t-square

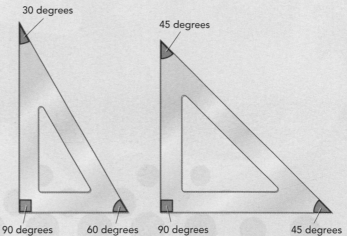

30 degrees

45 degrees

90 degrees   60 degrees   90 degrees   45 degrees

Other equipment

Name	Use
Drawing board clips or masking tape	Used to hold the drawing to the board
Set square	Helps you draw a line at standard drawing angles (30, 45 and 60 degrees) as well as 90 degrees
Pencil	Used for drawing on paper and available in various grades of hardness
Drafting eraser	Lifts pencil marks cleanly off the paper
Scale ruler	Allows you to draw objects smaller than their actual size

! Remember

We normally draw buildings at a **scale** of 1:100 for elevations and 1:50 for more details.

◎ Activity: Drawing scale lines

Draw the following outline of a house plan on your paper to a scale of 1:100.

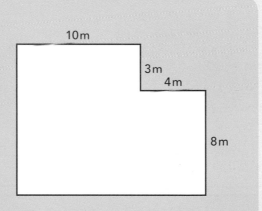

10m

3m
4m

8m

✔ Check

- Producing drawings involves the use of a lot of equipment

- Drawings are produced to different scales

L02 Setting up your drawing

Preparing a sheet of drawing paper

Any finished drawing needs the following added to it to make it a good means of communication:

- **Title** of the drawing

- What scale it is drawn to

- Who drew it

- Date

- Drawing number.

A drawing is normally set out like this:

✱ Key term

Title
The name given to something, in this case a drawing.

Border
A box that surrounds the drawing.

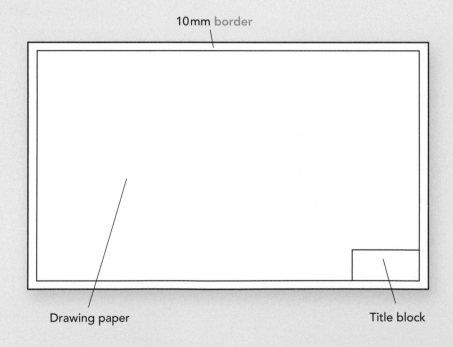

10mm border

Drawing paper

Title block

The title block which goes on the bottom right-hand corner would look like this:

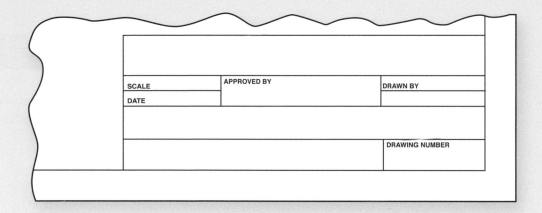

Activity: Draw a title block

Take one sheet of A3 paper and set it out with a border and title block.

Check

- Setting up a drawing requires a **border** and title block
- You should include the date and scale in your title block

L03 The construction drawing

The section

Sometimes a **section** is required through an object to show:

- How it is constructed

- Cross section details

- Thickness dimensions.

A section is normally drawn in a larger scale than the main drawing as it contains more details. Typical sections through certain elements are as follows.

Cavity wall

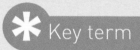

Key term

Section
A cut through an object to show the inside.

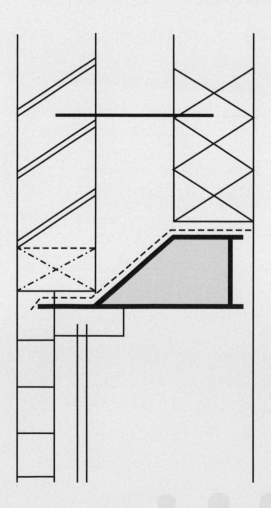

Ground floor and foundation

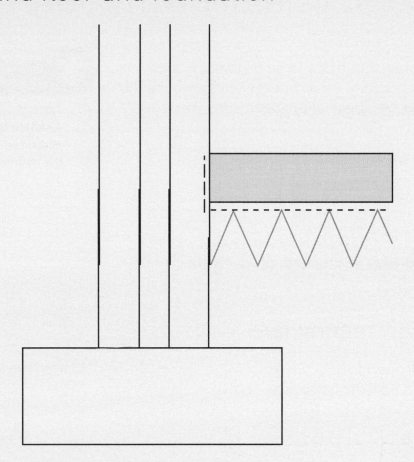

Activity: Drawing a cross section

Produce the following cross section by drawing it full size on a piece of A4 paper.

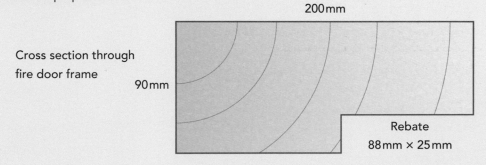

Cross section through fire door frame

200mm

90mm

Rebate
88mm × 25mm

Check

- Cross sections show what is inside an item

L04 Working responsibly with others

The production of a design, for example of a large building, would involve more than one person drawing. An architect would produce the initial design sketches and then get these approved by the client. A design team would involve:

- An engineer to design the structure

- An architectural technician to draw in the details

- A services engineer to design the services.

All of these people will have to work in a team to produce the client's final design.

Your role as a design technician

Working within a team you will need to:

- Act responsibly

- Understand your weaknesses

- Know your strengths

- Know other team members' strengths

- Cooperate when required

- Work efficiently.

Key term

Technician
A skilled technical person who completes the main work.

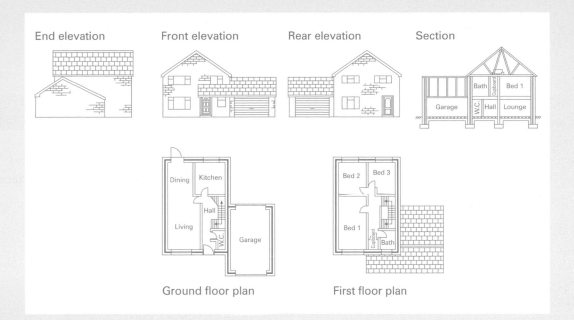

The photo shows the final design drawing of a house, which has been produced by the design team. The client can see what the shape and area of the house will look like.

Activity: Cooperation

You have just joined a new design team as a design technician. It is your first day at work. How will you make sure that others cooperate with you in a friendly and approachable way?

✔ Check

• Working in a team requires a number of different skills

• You need to take responsibility for your actions

! Remember

You will often have to take responsibility for any action you do.

 **Functional skills**

Communicating within a team will help develop your speaking and listening skills in **English**.

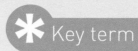
L05 Team working

Often you will be asked, or instructed, to undertake some tasks while working within a team drawing a design. In order to do this correctly here are some useful suggestions:

- Always be enthusiastic and willing to take on your duties – people will notice this

- Make sure you have an **open** and friendly approach to suggestions and advice from other team members

- Speak clearly and concisely to get your message across and don't interrupt people

- Listen carefully to any instructions and ask if you are not sure

- Ask questions at the right time and listen to the answers.

✳ Key term

Open
Behaving in a friendly manner with no barriers.

Functional skills

Speaking and listening will help you with your verbal skills in **English**.

Activity: Working in a team

With a partner, look at the foundation drawing (see *Ground floor and foundation* on page 45).

Draw the outline of the:

- Foundation

- The cavity wall

- The concrete floor.

Pass this over to your team member who will then have to complete:

- The drawing border

- The title block

- The shading in of each part.

Case study:
Dealing with situations

Joe is the architect's assistant on a design project for a new house. The architect has drawn the layout of the house and a general design of what it will look like. Joe has been asked to fill in the details to fit the outline.

Joe produces what he feels is a reasonable final design for the house and presents it to the architect for comment.

'No, that's not what I wanted,' said the architect.

Joe now realises that he should have made a rough outline drawing, then got this approved before moving on to a more detailed drawing.

- How would you deal with a situation like this?

✔ Check

- When working in a team, you must understand that what you do has an affect on the other team members

Assessment overview

While working through this unit, you will have prepared for completing the following tasks:

○	1.1	Select the drawing equipment required to produce a construction drawing	Pages 40–41
○	2.1	Create the border and a completed title panel for a construction drawing	Pages 42–43
○	3.1	Draw a vertical section through a cavity wall, the strip foundation to the wall and a concrete ground floor	Pages 44–45
○	4.1	Demonstrate good team working skills by working responsibly and cooperatively	Pages 46–47
○	5.1	Follow instructions when working with others	Pages 48–49
○	5.2	Communicate appropriately with others	Pages 48–49

Assignment tips

- Keep your hands washed and clean so that you will not get marks on your drawing.

- Use a sharp pencil to produce sharper lines.

- Use the correct eraser when removing lines.

- Don't press too hard when drawing.

- Use a border and title block.

DEVELOPING BRICKLAYING SKILLS

The craft of bricklaying dates back over 6000 years and since then the basic bricklaying tools used have hardly changed. This means that the skills and methods you will learn have also not changed and are similar today to how bricklayers worked centuries ago.

In this unit you will learn:

- About the hand tools used in basic bricklaying processes

- About the materials used in basic bricklaying processes

- About the Personal Protective Equipment (PPE) used in basic bricklaying processes

- To apply safe working practices to produce half-brick walling

- To work responsibly with others

- To ask for and respond to guidance when working as part of a team.

Why is it important for brick walls to be built straight?

LO1 LO2 Brickwork hand tools & materials

Handtools

The following table shows the variety of different hand tools used by bricklayers. The tool in each photograph is followed by an explanation of what it is used for.

✳ Key terms

Mortar
A mixture of sand, cement and water used to rest bricks on so they are level.

Plumb
Straight and not leaning in any way.

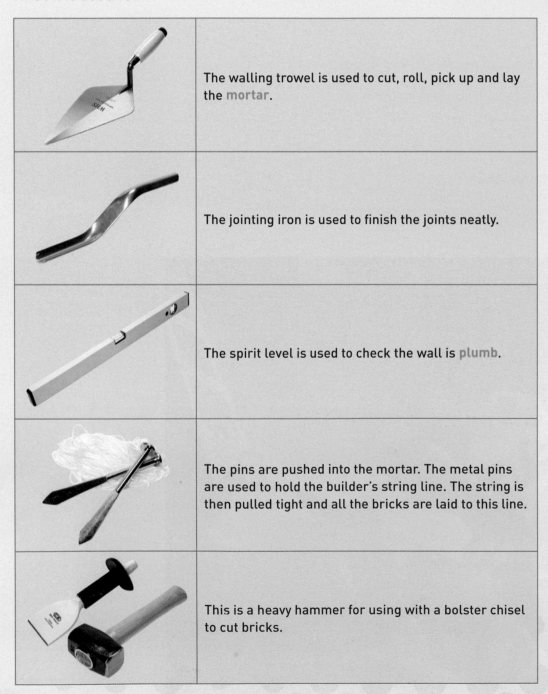

	The walling trowel is used to cut, roll, pick up and lay the mortar.
	The jointing iron is used to finish the joints neatly.
	The spirit level is used to check the wall is plumb.
	The pins are pushed into the mortar. The metal pins are used to hold the builder's string line. The string is then pulled tight and all the bricks are laid to this line.
	This is a heavy hammer for using with a bolster chisel to cut bricks.

Activity: What tool?

- What are a builder's string line and pins used for?

- What is a walling trowel used for?

Materials used in bricklaying

Bricks are made from clay and are fired in a brick oven, so they become hard and strong. Brick manufacturers may use different coloured clays and it is this that gives the great variety of coloured bricks. Today, some types of brick may contain recycled materials, such as glass and plastic. This saves on clay and also recycles materials.

The mortar used in training bricklayers does not contain cement, which makes it easier to take the wall down when you have finished. The mortar you will be using contains a mixture of building sand and a material called lime. See the next section on the precautions to take when working with lime-based mortar.

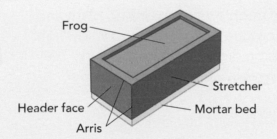

Activity: Building a stopped end

Without using mortar, use bricks and a cut half brick to build a stopped end that is plumb. Remember to use your spirit level.

✔ Check

- Bricks are made from clay and are fired in a brick oven

- Your spirit level makes sure that the walls you build are straight

L03 Personal Protective Equipment

When carrying out basic bricklaying processes make sure you are using the correct Personal Protective Equipment (PPE).

Hard hat

Wear a hard hat on all jobs where there could be a risk of any materials falling on your head or of you banging your head against an obstruction.

Eye protection

Eye protection must be worn when you are undertaking tasks where dust or debris could enter your eyes and cause an injury. Safety glasses or safety goggles are available depending upon the task you are doing and the level of protection that is required.

High-visibility jacket

A **high-visibility** jacket is a yellow (or orange) jacket that is worn over your clothes or overalls. It is 'high' in colour, so people who are driving plant and equipment can see you and avoid you. They are especially good in low light conditions and often have reflective strips on them for use in the dark. They are also available as a bib for more comfortable wearing in hot weather.

Safety boots

Safety boots need to be worn on sites to prevent any injury to your toes when handling materials. There is also a variety of other safety footwear, such as shoes and trainers available.

Hand barrier cream

Because a bricklayer works with many different products that often contain chemicals it is a good idea to use a barrier cream. This is rubbed into your hands before you start work and acts as a 'barrier' to prevent products from coming into contact with your skin.

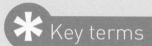

Key terms

High-visibility
Uses a contrast colour, usually yellow or orange, that is visible to the eye.

Dry bonding
Laying bricks in position with no mortar to check the length of the wall.

Gauge rod
A piece of straight timber marked out with saw cuts for each brick course.

! Remember

Check and wear the correct Personal Protective Equipment.

◎ Activity: What is PPE?

Make a list of all the PPE you may need when laying bricks.

Basic bricklaying processes

Building your courses

Before you start bricklaying you will have to prepare the area where you are going to build your wall. You will have to safely collect, lift and stack your bricks in your work area. Do not carry too many bricks at the same time.

The rows of bricks are called courses and it is very important that the first course of bricks is laid correctly. This is because every other course of bricks will rest on top of the first one.

Activity: Getting your course plumb

Draw around a coin and make two circles. In each circle draw the lines you see in a spirit level bubble. Next, draw where the bubble is when:

- The spirit level is in a plumb position

- The spirit level is NOT in a plumb position.

Label your drawings 'plumb' and 'not plumb'.

Dry bonding

Dry bonding is laying bricks in position with no mortar to see if the wall is the correct length.

If you are happy with the length of the dry bond wall, then you can start laying your bricks to the same length. You should start at the two ends of the wall and build upwards. Remember to check your wall at regular intervals with the spirit level so it is plumb. A **gauge rod** can be used to check the height of the brick courses as you build. Your tutor will show you how to do this.

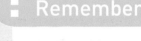

✔ Check

- PPE is important and must be worn (this is the law)

- A row of bricks is called a course of bricks

- Stack your bricks correctly so they will not fall over

! Remember

Lift and place down the bricks safely and correctly as your tutor has shown you.

! Remember

The gap between each brick should be 1 cm (10 mm).

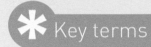

⬤L04 Safe working practices to produce half-brick walling

Safe working practices

It is very important that you follow the safety rules, regulations and guidelines for working safely in bricklaying. *Unit 10 Health and safety and welfare in construction* covers safe working practice in construction areas. You should have already completed, or be working on, this unit before starting your bricklaying.

There are certain hazards and protective measures that you will need to know about when you are bricklaying.

Safe maintenance, use and storage of tools

You will need to make sure that the tools that you are using are:

- Checked regularly for defects

- Replaced if damaged.

Case study:
Maintenance of equipment

Joe was a self-employed bricklayer and after qualifying at college had set up his own bricklaying firm. He had ten years of experience in building new houses. One day on site some bricks fell from the scaffolding and Joe received serious head injuries, even though he was wearing his hard hat. His hard hat did not protect him – it was over ten years old. Joe was not able to work for three months. Fortunately, he has now returned to work, but he cannot work as quickly as he used to.

Discuss the following questions in a group:

- Whose fault do you think this accident was?
- How do you think this accident could have been prevented?
- How do you think Joe's accident changed his family life?

Joe is a good example of what can happen if you don't take care of your Personal Protective Equipment (PPE).

✻ Key terms

Buttering
Placing mortar on the end of the brick or header.

Half-brick wall
This is a wall which has a depth or thickness of only half a brick.

Half-brick walling

Activity: Build a wall in stretcher bond

For this activity your wall should be seven bricks long and five courses high.

- Lay your first bed of mortar in position about 1 cm thick.
- Place the first and seventh bricks the same distance apart as when you laid them out with dry bond.
- Place the other five bricks in between these.
- When laying bricks you need to put mortar on the end of the brick or header. This is called buttering. There are various ways of doing this and your tutor will show you how.
- Straighten up your first course with a straight edge. Make sure your bricks are level.
- Remember to use a cul half brick at either end on the second course, and then whole bricks.
- Start your third course.

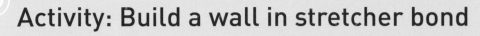

Remember

Don't put too much mortar on the header as the joint only needs to be 1 cm thick.

Activity: Half-brick wall

- What uses do you think there are for a half-brick wall?

Functional skills

Working out the measurements of your wall will help you develop your skills in mathematics.

Check

- Building a brick wall is an important skill in construction
- You don't need to put too much mortar on your header

L05 L06 Responsibility & teamwork

Behaviour

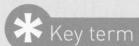 Key term

Behaviour
The way in which you act – your conduct or manner.

When you are working within the workshop always make sure that you act sensibly and show the correct **behaviour**. This means:

- Recognising your own and others' strengths and skills

- Cooperating with others

- Tidying up as you work.

 Case study:

Working responsibly

Ryan and Mitchell had attended the same school and were good friends. They had both just completed their Level 2 NVQ in bricklaying at college. Both were working on a new build housing project. The housing project was behind schedule and both of them had been doing lots of overtime. While waiting for more bricks to be loaded out by the forklift onto the second lift of scaffolding, an argument started between them. A scuffle started and Ryan fell from the scaffold. The brick guard had been removed to load out the bricks. Ryan suffered severe spinal injuries and is now confined to a wheelchair. He can no longer work as a bricklayer and is retraining as an IT technician.

Think about the following questions:

- Whose fault was the accident?

- What could have been done to prevent the accident?

- How do you think Ryan's life has changed?

Maintaining a clean and tidy work environment

Good housekeeping is essential in order to maintain a safe working environment. This means:

- Not leaving any sharp objects on the floor or working surface
- Removing all packaging which is a fire risk
- Cleaning work areas from debris and obstacles
- Placing all tools and equipment into tool bags and storing after use
- Disposing of any waste safely.

Following advice and guidance

When working in bricklaying it is important you follow instructions. This may be from your tutor or teacher, or possibly just a safety sign telling you what to do. From time to time your tutor will call all of you in to:

- Demonstrate a particular technique
- Give instructions on the next part of your practical
- Advise on any mistakes that you are making.

You will need to take this advice and guidance on board and act upon it, asking any further questions if you do not understand what is required.

Communicating with others

Good communication skills are important when bricklaying. Good communication involves:

- Listening carefully
- Asking questions
- Speaking clearly
- Following instructions.

Activity: Communication

- How could not asking a question cause someone to have an accident?
- How could not speaking clearly cause a construction accident?

Check

- Tidy up as you work
- Always follow instructions

Functional skills

Listening carefully and speaking clearly are examples of good English skills.

Remember

Your attitude is very important. You need to be:

- Enthusiastic
- Approachable.

Assessment overview

While working through this unit, you will have prepared for completing the following tasks:

○	1.1	List and describe appropriate hand tools to be used in basic bricklaying processes	Pages 52–53
○	2.1	List and describe appropriate materials to be used in basic bricklaying processes	Pages 52–53
○	3.1	List and describe appropriate PPE to be used in basic bricklaying processes	Pages 54–55
○	4.1	Select and use hand tools safely to lay bricks in stretcher bond, minimum seven bricks in length, minimum five courses high, with one stopped end	Pages 56–57
○	5.1	Maintain a clean and tidy work environment	Page 58
○	5.2	Work responsibly in the workshop	Page 58
○	6.1	Follow instructions when working with others	Page 59

Assignment tips

- When using lime-based mortar, try not to touch the mortar with your hands. Use the bricklaying trowel.

- When lifting and lowering bricks, keep your legs slightly apart and your back straight.

- Check for plumb with the spirit level regularly as you build and also check the brick courses with a gauge rod.

- Keep your building string line tight, so it does not sag.

- Your mortar bed should be about 1 cm thick; bricklaying trainees often start laying the mortar bed too thick to start with.

- To learn the craft of bricklaying you will need to practise a lot.

DEVELOPING CARPENTRY SKILLS

This unit introduces the hand tools, materials, Personal Protective Equipment (PPE) and skills used in carpentry. You will have the opportunity to produce a carpentry item.

Carpenters work on-site and perform a great variety of different tasks, such as fitting doors and door frames, door architraves, locks, skirting boards, staircases and timber roofs. Carpenters may also fit bedroom furniture and cupboards. This huge range of different carpentry tasks makes a carpenter's work both varied and interesting.

In this unit you will learn:

- About the hand tools used in basic carpentry processes

- About the materials used in basic carpentry processes

- About the Personal Protective Equipment (PPE) used in basic carpentry processes

- To apply safe working practices to produce a carpentry item

- To work responsibly with others

- How to ask for and respond to guidance when working as part of a team.

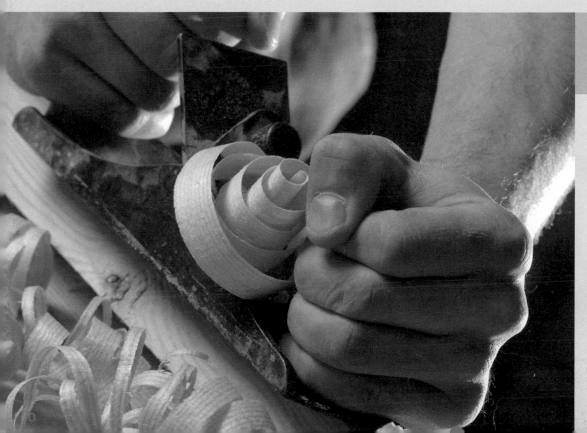

What things in a typical house might a carpenter be responsible for making and assembling?

L01 Carpentry hand tools

The pictures below show the variety of different hand tools used by carpenters. The tools are grouped by what they are used for.

Marking out tools	
	A marking gauge is used to mark a single line along the timber edge.
	A tri-square can be used to set out lines accurately on timber. You can also check that pieces of timber are at 90° (or right angles) on the corners.
	A sliding bevel can be used to set out angled lines on timber and also measure various angles.

Cutting tools	
	A tenon saw is used to cut woodwork joints or small pieces of timber.
	A bevel-edged chisel can be used to cut out a housing joint.
	A smoothing plane can be used to plane sawn timber.

Fixing tools

	A nail punch is used to punch nails below the surface of the timber.
	A screwdriver is used for driving in and removing screws.
	A mallet can be used for gently tapping joints together or hitting chisels when making housing joints.

Drilling tools

	The wheel brace hand drill is used with a drill bit to drill holes in timber.
	A bradawl is used to start or form small holes in timber.

Activity: Identifying tools

From the list of carpentry hand tools used for marking out, identify the tools that have a moving part.

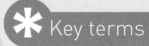

Key terms

Bevel
A slope or angle.

Plane
A tool used to shave timber to make it smooth.

Check

- Carpentry hand tools can be sorted into different groups by what they are used for: marking out, cutting, fixing and drilling
- A tenon saw is used to cut woodwork joints or small pieces of timber

L02 Carpentry materials

Softwood timber

Softwood timber comes from softwood trees. In general, softwood trees:

- Are evergreen trees which do not lose their needle-like leaves in winter

- Produce pine cones

- Are fast growing.

The timber is easy to work with and light in colour. It is used for making roofs, staircases, floors and doors.

Plywood

This material is man-made and is formed by glueing thin sheets of **veneer** together to form a board. The veneers are cut from trees. Each veneer sheet is glued over another and this gives the plywood strength.

Nails

There are many different types of nails. Oval nails (below left) are oval in shape and are less likely to split the timber when used. Panel pins (pictured right) are small round nails and can be used to fix small **mouldings**.

* Key terms

Veneer
A very thin sheet of timber.

Moulding
A piece of timber specially shaped to fit around something.

◎ Activity: Name the material

Name a type of sheet material and two different types of nails used to fix timber.

Woodscrews

Woodscrews come in different lengths and sizes. They are very useful because they can easily be removed from timber. Woodscrews are used to fix metal hinges to doors and door frames. The head of the screw can vary. A slotted screw head is shown on the left below, and a posi-drive screw head is shown on the right. The correct screwdriver must be used with certain woodscrews.

Hinges

There are many different types of hinge and they can be made from steel or brass. They are used by carpenters to hang doors in buildings: one flap attaches to the door, the other to the door frame, screwed through pre-drilled holes with woodscrews.

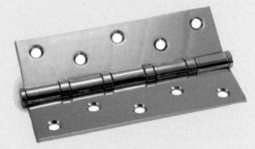

Activity: Drawing a hinge

Make a drawing of an open hinge and a closed hinge.

Check

- It is important to have the correct material for the job you are doing

- Knowing about the features of materials lets you choose the right material to use

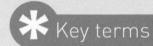
L04 Basic carpentry processes

Planing timber

Carpenters have a variety of planes for different **planing** tasks. Planing timber gives the wood a smooth finish and makes it slightly smaller, which helps to make it fit with other pieces of wood. When you plane timber you produce long thin rolls of timber called wood shavings. Carpenters plane timber to:

- Remove the rough sawn surface and make it smooth

- Make the timber straight and square so that it is easier to use.

Activity: Make a photograph frame

In this activity you will be making a photograph frame with hinges.

Your tutor will give you:

- Softwood timber that has had a part cut out to hold the back of the frame

- A dimension sheet to work out lengths.

1. Correctly cut your timber to the right length using the dimension sheet.

2. On the longest side mark out and cut the hinge for the photograph frame piece and fix the hinge using screws.

3. Use **mitre** joints at the corners and produce a rectangular frame.

4. Pin and glue together. You may need to use a **vice** to help you.

5. When the photograph frame has dried, remove any visible glue marks by lightly sanding.

6. Cut a piece of plywood for the back insert of the frame so it fits within the recess that you have formed.

7. Use metal fixings to hold the back.

8. Using the hardboard template, shape a frame stand and attach this, pre-drilling the holes.

9. Fix the frame stand to the hinge using woodscrews.

✳ Key terms

Planing
Shaving timber to make it smooth.

Mitre
An angled cut in a piece of timber.

Vice
A moving clamp fixed to the side of a bench for holding timber tightly.

❗ Remember

Check the stand works!

Activity: Cutting out a housing joint

Cut out a housing joint for a hinge plate. Remember to listen to your tutor's instructions on how to do this and stay safe.

Check

- Carpenters plane timber to give their work a smooth finish, to make the timber smaller and to make the timber square

- When working with wood, pay attention to your own and other's safety

Functional skills

Listening to instructions will help you develop your skills in English.

L03 L05 L06 Working safely & responsibly

Unit 10 Health and safety and welfare in construction introduced you to health and safety, welfare issues and safe working practice in construction. You may have already completed, or be working on, this unit before starting your carpentry. Refresh your memory on pages 11–26.

As a reminder, all Personal Protective Equipment (**PPE**) should be:

- Used correctly

- Replaced if damaged.

◎ ## Activity: Types of protective equipment

Name three pieces of PPE used in carpentry and write a description of each one. Then choose a partner and talk about the possible **hazards** that each piece of PPE might be used against.

When working in carpentry you need to:

- Maintain a clean and tidy work area. This means tidying away rubbish, for example wood shavings and sawdust, so that they do not build up and cause a hazard. All waste timber should be cleared away and not left lying around because someone could trip over it.

- Work responsibly in the workshop. This means using, keeping and storing your carpentry tools safely at all times. Place your tools neatly near you when you are working, so they do not get in your way or cause an accident. When in the workshop you also need to behave in a responsible manner. This means following the rules and safety signs in the carpentry workshop, for example no running or messing about, and wearing your safety boots.

✳ ## Key term

PPE
Personal Protective Equipment, designed to keep you safe while you work.

Hazards
A danger, such as a loose hammer head.

Functional skills

Talking about things with a partner or a workmate could help you develop your skills in English.

! ## Remember

- Clean up as you are doing your carpentry work.
- Messing about in the carpentry area can cause an accident.

Case study:

Kevin's workplace accident

Kevin has been working for a carpentry firm for ten months on a large housing project. When he returned from lunch, at the site entrance, Kevin started helping with unloading some long timber joists which had just been delivered. Everyone on the site was helping and they were in a hurry as the driver was parked in a bus lane and was worried about getting a parking ticket.

While collecting one of the timber joists from the vehicle by himself, Kevin received a deep splinter in his right hand. He was taken to hospital and needed a minor operation to remove the splinter. Kevin was off work for two weeks while his hand healed.

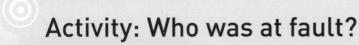

Activity: Who was at fault?

Whose fault was the accident? Discuss this with a partner. How do you think Kevin could have avoided this accident?

Check

- Always use your Personal Protective Equipment
- Do not rush
- When carrying and lifting get help if you think you need it

Assessment overview

While working through this unit, you will have prepared for completing the following assessment tasks:

○	1.1	List and describe appropriate hand tools to be used in basic carpentry processes	Pages 62–63
○	2.1	List and describe appropriate materials to be used in basic carpentry processes	Pages 64–65
○	3.1	List and describe appropriate PPE to be used in basic carpentry processes	Page 68
○	4.1	Select and use hand tools safely to make a carpentry item (photograph display item with hinge(s)) in an acceptable time	Pages 66–77
○	5.1	Maintain a clean and tidy work environment	Pages 68–69
○	5.2	Work responsibly in the workshop	Pages 68–69
○	6.1	Follow instructions when working with others	Page 69
○	6.2	Communicate appropriately with others	Page 69

Assignment tips

- When cutting out a woodwork joint with a saw or a chisel, always follow the marked line down the side of the wood that you will throw away.

- When using a plane, make sure that the blade is sticking out only a very small amount under the plane.

- When using small nails, blunt the pointed end of the nail with a hammer before using it. This will make your timber less likely to split when nailing it.

- Do not forget to punch your nails in with a nail punch.

- If the face of your hammer is dirty it will slip off the top of the nail when you strike it. Clean the face of the hammer with abrasive paper to prevent this happening.

- Take your time when measuring your timber, which means checking your measurement. ('Measure twice, cut once.')

DEVELOPING JOINERY SKILLS

The modern home has many different items of manufactured joinery used in its construction. These cover the stairs, handrails, doors, frames, skirting boards, shelving, windows and architraves. Timber products are used to make the home look attractive and pleasing to a potential buyer.

There is a great deal of skill involved in working with timber, as it cannot be repaired once it has been cut, and so great care must be taken with the marking out work involved.

In this unit you will learn:

- About the hand tools used in basic joinery processes
- About the materials and fixings used in basic joinery processes
- About the Personal Protective Equipment (PPE) used in basic joinery processes
- To apply safe working practices to produce a joinery product
- To work responsibly with others
- To ask for and respond to guidance when working as part of a team.

How difficult do you think it would be to make this staircase?

L01 Joinery hand tools

The pictures below show the variety of different hand tools used in basic joinery processes.

 Key term

Paring
Using a chisel by hand to carefully remove sections of timber.

	A steel ruler is stronger than a plastic one so it won't wear or break and can be used for the accurate marking out required for joinery.
	A sliding bevel has an adjustable blade on it so that angles can be copied and marked out.
	This is a joiner's pencil. It isn't round, but flat, so when used outside it won't blow away or roll off a surface.
	This is a marking/mortice gauge. It is used to mark out special joints in timber that have a tongue and a slot which need to be the same width. Using this gauge produces the correct width.
	A mallet is a type of hammer made from timber. It has a large head which is used with a chisel to cut out slots from timber.
	A tenon saw is used for cutting dovetail joints in joinery products as it has a very fine blade and is very accurate.
	A chisel is a hand tool that is used to cut across the grain to remove timber. It is used with a mallet or on its own by paring.

	A nail punch is used to push a nail below the finished surface so that it can be filled and decorated.
	This is a claw hammer. It is named after the claw that is used to remove nails from timber.
	This wheel brace is a tool that holds cutting drill bits in a jaw and is then driven round by hand so that it drills a hole into the timber. Various size holes can then be cut by hand.
	There are two types of screwdriver – the flat head and the cross head. They fasten screws into timber and are used to secure fixings.
	This is a smaller plane that has a wider blade for smoothing down timber sections prior to fitting together.

Activity: Tool identification

Identify the following tools:

- This tool has a handle and a steel blade and is used on posi-drive heads

- This tool had its angle adjusted and was transferred over to the new piece of timber to mark it

- This tool was used to carefully cut the dovetails.

✔ Check

- A large number of hand tools are required for joinery production

- Two types of screwdriver are used in the UK

L02 Joinery materials & fixings

The materials and fixings used in basic joinery processes are listed below.

Softwood timber

There are two types of timber grown in the world – softwood and hardwood.

Softwood timber is mainly used to make joinery items for house construction in the UK.

PVA glue

PVA is glue that is white in colour and is used to fasten pieces of timber together. It is normally applied to both surfaces, which are then clamped together under pressure until the glue has set. The glue forms a strong bond between the two timber surfaces.

Oval nails

An oval nail has an oval head shape rather than a round shape. When it is used on timber lengthways its head can be driven below the surface and the grain closes over, leaving a neat hole that can be filled.

Key term

PVA
Polyvinyl acetate glue.

Pilot hole
A small hole used to make sure that a screw is fixed in the correct place.

Panel pins

These are very small thin nails that are used to fasten sheet materials to timber frames.

Woodscrews

Woodscrews have a thread that draws the screw into the timber. It is always best to drill a **pilot hole** before you drive home the screw.

Activity: What type of fixing?

What type of fixing is required to fix a hinge and ply back to a picture frame?

✔ Check

- The type of fixing used will depend upon the item being secured

L03 Personal Protective Equipment (PPE)

Eyes

Your eyes have to be protected when machining joinery. A saw will create flying particles and dust. You would need to wear safety **glasses** or **goggles**.

Feet

The hazard with carrying heavy objects comes when they slip. This can cause damage to the bones in the feet, as well as causing crush injuries should a machine run over the foot. Injuries from nails left in timber on the ground can puncture the base of the foot. Safety boots are worn because:

- They have steel toecaps to protect the toes

- They have steel plates in the base to prevent nails piercing the foot

- They have an ankle guard to support the ankle

- They have a tough tread sole to prevent slipping.

Nose

The nose, and your respiratory (breathing) system, is prone to the effects of the sawdust produced when working with timber. This can cause some discomfort and irritation, and long-term exposure can cause lung disease. To avoid these problems a dust mask should be worn over the nose and mouth.

Ears

Our ears are very sensitive. They can be damaged with exposure to loud noises over a period of time. To prevent this, ear protectors are worn that reduce the volume of noise entering the ear.

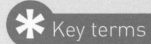

✳ Key terms

Glasses
These look like normal glasses and fit over the ears and nose to stay in place.

Goggles
These cover the whole of the eyes with a flexible plastic shield.

Head

Head protection is normally worn to:

- Prevent harm from any object that you bang into or which falls onto your head

- Set a good example on a site.

A safety helmet should be in-date and free from any cracks or dents.

Case study:

Which PPE?

Ben is working with softwood timber and is making and fixing a door frame. He has used a 110v planer to trim down the frame and is nailing the frame into the opening in the brickwork. After fitting the frame Ben secures the architraves using oval nails and hangs the door on its hinges.

- Identify which PPE Ben should wear when undertaking this task.

✔ Check

- PPE should be used as a last resort to protect the worker

- Each joinery task should be assessed for PPE correct use

L03 Basic joints

The housing joint

A housing **joint** is a slot cut across one section of timber which is the same width as the other part of the joint that fits into the slot to form a tight joint. This would normally then be glued together.

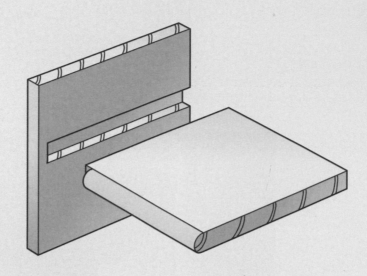

The tee halving joint

This forms the shape of a 'T' and uses a housing joint on one part and a tongue on the other half of the joint. These fit together so that they are the depth of half of the timber section and finish flush both sides. They are then pushed together to form a joint which is normally glued together.

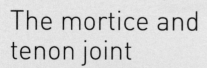

The mortice and tenon joint

The joint is in two pieces – a tenon and a mortice. The tenon (on the right) has been cut out, forming two shoulders on the end of the first piece. The second piece (on the left) has had a slot cut into the timber which is the same depth and the length of the tenon. The joint would be clamped and glued together.

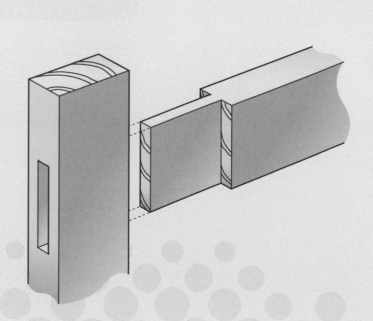

Activity: Practice joints

Functional skills

Marking out for joints will help you with measuring dimensions for mathematics.

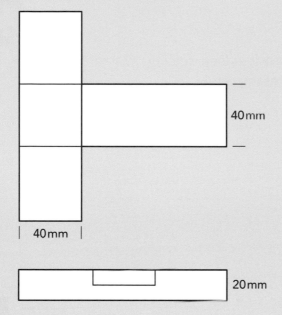

40mm

40mm

20mm

Your tutor will provide you with the correctly machined timber as per the thicknesses and widths in the pictures. Mark out and produce a sample joint, taking care to cut neatly so that it fits tightly with no gaps.

Check

- A range of joints are used in joinery products
- Care needs to be taken in measuring and cutting

L04 Safe working practices

Safe working practices

When you produce your **stool** joinery product (see activity opposite) you will have to carry out the following processes safely:

- Cutting timber with a saw

- Using a sharp chisel

- Using a marking gauge

- Sanding down your work

- Wearing the correct PPE

- Safely storing your hand tools

- Cleaning away any waste materials.

Good **housekeeping** and working next to your colleagues must be carefully undertaken so nothing can harm any other person.

Keeping your work area clear of tools you are not using will help:

- Prevent damage to the tools if they roll off the bench

- Prevent injury due to cuts from chisels and knives lying on the bench

- Maintain an attitude of keeping work clean, tidy and safe.

If you start as you mean to go on in a career in joinery, the safe working practices that you learn now will stand you in good stead when you start on a construction site. If you work safely then others will follow your good example.

Guidance

Your tutor will issue you with guidance from time to time on:

- Issuing the drawing for the stool

- Marking your setting out before you cut

- Helping with difficult details

- Final marking and grading of work.

Safe maintenance

You should always maintain a safe and tidy work area so no injuries can occur. To do this you will need to:

- Keep all loose materials stored in racks
- Keep shavings in a bin with a lid and clear them away at the end of the day
- Avoid any trip hazards on the workshop floor.

Storage of tools and equipment

When you are working on your stool you will need to ensure that your tools are well maintained by:

- Making sure that they are sharp
- Making sure they are properly stored in the workbench when not in use
- Making sure they are put away at the end of each practical session
- Reporting any defects or damage.

Activity: Joinery task

Your tutor will give out the detailed drawing for a stool that you are going to make using some of the joints that you have practised. You will need to:

- Read and understand the details required
- Ask any questions before you start to make sure that you know what you are doing
- Follow any instructions you are given
- Help colleagues with any problems.

Activity: Working safely

Read the section opposite on working safely a few times to make sure you are comfortable with this before you start working.

Remember

Measure twice, cut once.

Functional skills

Reading and understanding the drawing will help your skills in **English**.

Check

- Making a timber product takes skill and time
- Working safely is very important

L05 L06 Working responsibly with others

Behaviour

When you are working within the workshop always make sure that you act sensibly and show the correct **behaviour**. For example, carrying a chisel from the cupboard to your workbench has to be done correctly to avoid an accident. You are expected to behave as if you were working in an actual joinery shop making products for customers and clients.

You will be working in teams of two on a workbench. This will involve:

- Sharing tools
- Sharing ideas and information
- Talking and discussing progress
- Cleaning up as you go
- Sharing design drawings.

You will need to get on with your colleague so you both complete the task well, for example the stool in the joinery task. To do this you will need to:

- Individually bag or label your work
- Work each side and end of a workbench
- Motivate and encourage each other.

> **✳ Key term**
>
> **Behaviour**
> The way in which you act – your conduct or manner.

📁 Case study:
Working responsibly

Steve and Michael were sharing a workbench. They had been working on practicals all day and had not bothered to clear up any waste materials, put away tools they had not used or swept the floor. Paul, a colleague, on the next workbench, was walking past their work area when he tripped up over a hammer left on the floor and banged his head on the vice which was sticking out. Paul had a severe cut and an ambulance had to be called.

- What should Steve and Michael have done to avoid this accident?

Responding to guidance

Your tutor will from time to time ask you how you are getting on and offer guidance, advice and instructions when you are working. This will be:

- To show you how to perform better

- To make you produce better quality work

- To correct mistakes so you learn from them

- To help with any problems you have.

Communication skills

Your tutor will talk to you and discuss your progress with the joinery task. You will need to:

- Listen carefully to what is said

- Practise any techniques shown

- Ask for work to be checked.

Functional skills

Communicating by talking will help you with your functional skills in English.

Activity: Working together

Keep a note of the following as you work through the joinery task with your colleague:

- When did I need the help of my colleague?

- What difficulties did we have?

- How would I do things differently?

Check

- Producing a final product often means working with others

- You need to behave in the correct way to make sure the work gets done well

Assessment overview

While working through this unit, you will have prepared for completing the following tasks:

○	1.1	List and describe appropriate hand tools to be used in basic joinery processes	Pages 72–73
○	2.1	List and describe appropriate materials and fixings to be used in basic joinery processes	Pages 74–75
○	3.1	List and describe appropriate PPE to be used in basic joinery processes	Pages 76–79
○	4.1	Select and use hand tools safely to produce a stool in an acceptable time	Pages 80–81
○	5.1	Maintain a clean and tidy work environment	Page 82
○	5.2	Work responsibly in the workshop	Page 82
○	6.1	Follow instructions when working with others	Page 83
○	6.2	Communicate appropriately with others	Page 83

Assignment tips

- Measure twice and cut once. This will save you from making lots of mistakes.

- Re-check dimensions transferred from a drawing.

- Keep your workbench tidy and free from hazards.

- Wear appropriate PPE as required.

- Be very careful when carrying a chisel – make sure you hold it as your tutor has shown you.

- Always wash your hands as it will keep your work clean.

- Use a sharp pencil to mark out your work.

DEVELOPING CONSTRUCTION PAINTING SKILLS

Every home in the UK has been painted at some time during its life. New homes have to be decorated so they look smart and attractive to potential buyers. Painting involves the application of wet paint onto a surface, which then dries and forms a clean, colourful coating to walls, ceilings, doors, frames, skirtings and architraves.

Paint is available in thousands of different colours. This is used to make attractive features on walls and to produce a pleasant space and environment to work, live and play in. Paint can be applied inside and outside of a building.

In this unit you will learn:

- About the hand tools and equipment used in basic painting tasks
- About the materials used in basic painting tasks
- About the Personal Protective Equipment (PPE) used in basic painting tasks
- To apply safe working practices to paint a flat wall area
- To work responsibly with others
- To ask for and respond to guidance when working as part of a team.

Is this painter using a correct method?

L01 Painting tools & equipment

The **hand tools** used in basic painting tasks are shown below and opposite.

Abrasive paper

This is a paper that is coated with sand, or manufactured abrasive, that is used to smooth down a surface before you apply any paint and to **key** a previously painted surface so that the new paint sticks.

Paint scrapers

This is a paint scraper, which is used to remove any imperfections found during the painting **preparation**. It is also used to remove old wallpaper coverings. Paint scrapers are available in different width blades.

Shave hook

The shave hook is used to remove old paint when using a heat gun or blow lamp to melt it first. The shave hook then peels off the paint down to the bare wood.

> ❗ **Remember**
>
> Removing old paint can be hazardous if it contains lead.

Filling knife

A filling knife is used to mix and apply fine surface filler into any surface defects. It is thinner than a paint scraper and is more flexible.

Paint kettle

Paint for commercial painting is normally delivered in large containers. To try and use this on a ladder would prove dangerous, so a paint kettle allows you to pour out a smaller amount and work with this. A paint kettle is traditionally made out of steel but modern ones are plastic.

Paint brush

Modern paint brushes are manufactured from artificial bristles that are held together with a metal band fixed to a brush handle. They are available in a variety of different widths for different tasks.

Paint roller

The paint roller has a soft absorbent foam tube that soaks up paint and applies it to the surface that you are covering. A pole can be fitted to the handle to extend your reach. A roller is a much quicker method of applying paint to large areas.

Paint roller tray

A paint roller tray is used with the roller. It has a deep end for the paint and a shallow end to run off excess. The paint roller tray is used to store a supply of paint and to ensure an even coat on the roller before it is applied to the surface you are painting.

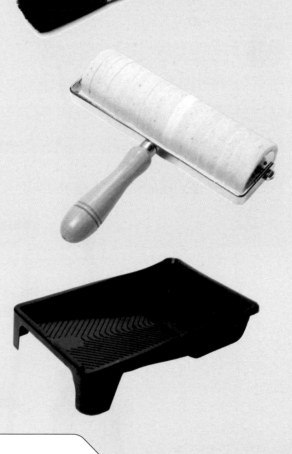

Check

- Preparing the surface is essential before painting
- It is important to have the right tools

ⓛⓞ¹ Equipment

The main equipment used in basic painting tasks is shown below and opposite.

Stepladders

Stepladders are a useful tool for painters to reach ceiling level if they are decorating plaster covings, mouldings or wall papering. They extend your normal height so you can reach up higher.

✳ Key terms

Stepladders
A series of steps with side rails hinged together to form a ladder.

Hop-up

This is a small low-level platform that you step onto to extend your reach. Hop-ups should be assessed for the risks in using them as they have no handrails.

Activity: Tool identification

- Read the passage below and then make a list of the painting tools you will need to carry out the job.

The room needs a new look brought to it. The client has found the colour required for the walls, ceiling and all timber surfaces. The height of the room is 2400 mms. One wall has existing wood chip paper on it which needs renewing.

- Why do you think you have to be especially careful when using stepladders?

✔ Check

- There are different tools required for preparation and painting
- The use of ladders and hop-ups must be assessed for risks

L02 Painting materials

Water-based paints

Most paints that are water-based are called emulsion paints. They dry by the water evaporating from the paint, leaving the paint bonded to the surface. Emulsion paints are produced in different surface finishes:

- Matt

- Vinyl silk

- Eggshell.

The vinyl silk and eggshell finishes are easier to wipe down, clean and maintain.

Paints are produced in many different colours.

Wood primer

Any bare and exposed timber requires a base **coat** to seal the timber. This prevents any natural chemicals from spoiling the final undercoat and gloss finishes. There are two types of basic primer:

- Aluminium – used on timber

- Alkaline primers – used on bricks, concrete, render.

Undercoat

This is the first coat of paint that starts to build up a protective thickness for the final gloss coat. It fills in the surface defects normally on the grain of the wood in preparation for the final coat. The primer should be sanded down before application of the undercoat which should be allowed to dry before rubbing down.

Gloss finish

This is a solvent, or oil-based, finish but which is often now replaced with a water-based gloss finish to make it more environmentally friendly. An eggshell finish has a matt rather than a high gloss finish. This is the final coat that forms a water-resistant and protective layer.

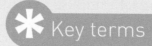

Key terms

Coat
One application of a paint finish which is left to dry.

Knotting
Used to seal knots in timber to prevent resin seeping out.

Other materials

We need other painting materials during preparation and for the cleaning of painting equipment. These are:

- White spirit – this is used to clean undercoat and gloss paint off brushes

- Turpentine – this has the same use as white spirit

- Knotting – this is dabbed over any knots in timber to seal them and prevent them leaking any natural resin which discolours the gloss finish

- Filler – this is mixed with water and applied, left to harden, then sanded smooth

- Paint stripper – this is a chemical that removes paint so it can be scraped off.

Activity: Preparation

You have been asked to paint some softwood skirting boards. Make a flow chart to show the process you will go through right through to the gloss finish application.

Functional skills

This activity will help you order your thoughts better and improve your writing skills in **English**.

Check

- Paints are available in a range of colours and finishes

- Gloss painting requires a number of different coats

L03 Personal Protective Equipment for painting

It is important to use Personal Protective Equipment (**PPE**) in basic painting tasks.

✳ Key term

PPE
Personal Protective Equipment.

Hands

Some workers may be prone to irritation from paint coming into contact with their skin. Thin polythene gloves can be worn to prevent this, but most water-based paints will wash off from the surface of the skin.

Activity: The hands

As we have seen, certain paints are made using water and other paints using chemicals. When you get paint on your hands, what action should you take?

Select the correct answer from this list:

- Ignore it as it will eventually wear off

- Read the manufacturer's instructions for removal

- Wash hands with a paint stripper

- Wipe hands on a used rag.

Eyes

If you are using chemical paint stripper, use safety glasses in case you flick any into your eyes, where it will burn.

Body

Overalls must be worn when painting. They are used to protect the body from paint splashes. The alternative, when it is a really messy painting task, is to use a disposable paper suit that can be thrown away when it is finished with. Overalls will need washing on a regular basis.

Activity: PPE selection

Identify when you would use a dust mask and gloves in painting tasks.

Activity: Hard hat!

The safety officer has told you off for not wearing a hard hat. You reply that you don't need to wear a hard hat as you are a painter. When would you need to wear one?

Case study:

Working safely

Anna and Tom were working as a team on the refurbishment of a series of houses that had kitchen extensions done on them. Tom was responsible for the preparation work on the walls and Anna was to coat the ceilings and walls with two coats of emulsion paint. After the painting work had dried they looked around the kitchen and realised that the paint from rollering had coated both their faces and hair and was starting to itch.

- What had they forgotten to do?

- How could this have been avoided?

Check

- PPE should be used as a last resort to protect the individual

- Each task should be assessed for PPE correct use

L04 Safe painting practices

Safe working

When you are painting you will need to ensure that you:

- Put down some dust sheets to prevent paint splashing on the floor
- Cover up any **fixtures and fittings**
- Don't put too much paint on the brush
- Wear eye protection if using a paint roller
- Use ladders correctly
- Store brushes correctly.

Basic painting

Do the following in this order:

- Prepare your wall area correctly by filling and sanding down any imperfections
- Emulsion paint the wall area, applying two coats of paint to 2 m²
- Carefully cut in to the timber moulding that is on the wall.

Painting preparation

To produce a quality finish you will need to be very careful with the preparation work. This will involve:

- Mixing and filling any holes or indentations
- Carrying out plaster repairs if holes are large
- Sanding down and refilling if required
- Removing and washing down to remove all the dust
- Making sure that the paint is mixed correctly
- Allowing the correct drying time between coats of paint.

Painting preparation of household items

Many household items will already be painted using gloss paint. The items can be stripped or rubbed down with fine sandpaper to receive an undercoat and a final coat of gloss paint.

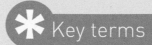

✱ Key terms

Fixtures and fittings
For example, a lamp, light or furniture.

Cutting in
Finishing correctly where one paint touches another of a different colour.

Load
Putting paint onto the brush to apply it to a surface.

Cutting in

Cutting in is a skill that is going to take time to learn. To do this properly you will need:

- A good clean brush of the correct width

- To make sure that the edge you are working to is straight and free from snags

- To **load** the brush with enough paint

- To make sure you work with a steady and flowing motion

- Continue until the brush needs filling again.

This technique will prevent the paint running over onto the other surfaces, and spoiling your clean line.

Activity: Practise your painting

Fill your paint brush and draw out the first letter of your first name on a plain wall. Fill in the letter by hand in a different colour, cutting in inside the letter to produce a clean finish.

Activity: Colour selection

Select a household item and then find a paint supplier's colour card. Try to match the existing paint colour with a recoat of this from the colour chart.

✔ Check

- Careful preparation is required to produce quality work

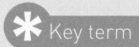

L05 L06 Working responsibly with others

Behaviour

You are going to be working with paint which is a wet material. It can:

* Damage clothing if spilt

* Damage footwear

* Cause harm if it enters the eyes

* Cause a slip if spilt on the floor

* Damage surfaces that are a different colour

* Damage fixtures and fittings.

Your behaviour therefore must be acceptable when you are working in the practical workshop applying paint. Fooling around when using such materials will not be accepted by your tutor and you may be asked to leave the workshop.

> ***** Key term
>
> **Behaviour**
> The way in which you act – your conduct or manner.

 Case study:

Working responsibly

Mohammed was working with a friend in the painting bay. He was painting his side of the wall, but his friend kept flicking his paint brush with paint over Mohammed's back for a laugh. Mohammed was getting pretty fed up with this, especially when his tutor asked why his back was covered in red splashes.

* Why is Mohammed's friend's behaviour not acceptable?

Responding to guidance

Your tutor will from time to time ask you how you are getting on and offer guidance, advice and instructions when you are working. This will be:

- To show you how to perform better

- To make you produce better quality work

- To correct mistakes so you learn from these

- To help with any problems you have.

Communication skills

Your tutor will talk to you and discuss your progress. The best way to help you is to use demonstration. Your tutor will actually show you how to apply the paint correctly and what techniques are required to produce quality work. You will need to:

- Listen carefully to what is said

- Practise the technique

- Ask for it to be checked.

Functional skills

Communicating by talking will help you with your functional skills in English.

Activity: Discussion prompts

If you are finding it difficult to talk with a team member, or your tutor, try writing down a list of queries that you can use the next time as a prompt for discussion.

Check

- Working sensibly with wet paint is good behaviour

- Taking time to listen means that communication is better

Assessment overview

While working through this unit, you will have prepared for completing the following tasks:

◯	1.1	List and describe appropriate hand tools and equipment to be used in the painting process	Pages 86–89
◯	2.1	List and describe appropriate materials to be used in the painting process	Pages 90–91
◯	3.1	List and describe appropriate Personal Protective Equipment to be used when painting	Pages 92–93
◯	4.1	Select and use hand tools safely to paint a flat wall area of 2 m²	Pages 94–95
◯	5.1	Maintain a clean and tidy work environment	Pages 96–97
◯	5.2	Work responsibly in the workshop	Pages 96–97
◯	6.1	Follow instructions when working with others	Pages 96–97
◯	6.2	Communicate appropriately with others	Pages 96–97

edexcel

Assignment tips

- Make sure that you prepare your area to paint well, before applying the emulsion paint.

- Always wash out brushes when you have finished painting before they dry solid.

- Take care when applying paint to avoid splashing.

- Always replace the lid on any tin after use.

DEVELOPING CONSTRUCTION DECORATING SKILLS

Decorating involves the application of wall coverings. These are wallpapers that provide an alternative to emulsion paint in covering plastered walls and ceilings. Wallpapers have been used for a number of years and are now available in thousands of patterns and textures. Decorating involves a great deal of skill to make the finished wall coverings look like quality products.

In this unit you will learn:

- About the hand tools used in basic decorating tasks

- About the materials used in basic decorating tasks

- About the Personal Protective Equipment (PPE) used in basic decorating tasks

- To apply safe working practices to produce a wallpapered wall

- To work responsibly with others

- To ask for and respond to guidance when working as part of a team.

How will the decorator get the wallpaper straight?

L01 Decorating tools & equipment

Tools

There are many hand tools that are used in preparing, hanging and finishing wallpaper tasks. Here is a selection that is commonly used.

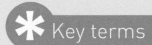

	A scraper is used with water to remove any old wallpaper.
	Filling knives are used to mix and fill any holes in walls.
	Wallpaper is pasted on a pasting table prior to hanging.
	A paste bucket holds the wallpaper paste.
	A paste brush is used to spread wallpaper paste.
	A paper-hanging brush is used to smooth out the wallpaper when it is hung.

	A caulker is used to fill any small gaps between timber and walls.
	Trimming knives are used to cut wallpaper against surfaces.
	Scissors or shears are used to trim or cut wallpaper length.
	A plumb bob and line is used to make sure wallpaper is vertical.
	A spirit level is used to set up a vertical line.

Activity: Tool identification

Answer the following questions on hand tools.

- Paper-hanging scissors are different from normal scissors. Why?

- What is a plumb bob used for?

- Why is there a clip on a pasting brush?

Check

- Paper-hanging equipment is often sharp and must be handled with care

- A variety of tools are required for pasting, cutting and finishing

LO2 Decorating materials

Wallpapers are used to cover large wall and ceiling areas and often have a variety of different coloured patterns on their facing side. A wall surface has to be prepared so the wallpaper will stick and hold until it has dried.

Lining wallpapers

A lining paper is used to:

- Cover up any minor imperfections in the wall before wallpapering

- Act as a clean surface that can be painted.

Lining paper is available in a range of different thicknesses. It is normally applied horizontally and then the wallpaper vertically.

Wallpapers

Wallpapers are produced in a variety of:

- Colours

- Patterns

- Textures

- Widths.

This gives a designer a range of different styles that can be applied to a room. Some wallpapers already have the adhesive on them and all you have to do is soak them in water. Most wallpapers have to be pasted with adhesive to bond them to a wall or ceiling. Wallpapers can be used without a **matching pattern**. They are easier to use as you don't waste so much and you don't have to be accurate in matching up each piece.

Pastes

Wallpaper paste is available as a dry product that has to be mixed with the required amount of water. You will need to measure in a bucket the correct amount of water before you add the paste and then stir until all the lumps and flakes have been dissolved into the paste.

Some tips when using wallpaper paste:

- Always mix enough for what you need

- Make sure you read the instructions

- Add the correct amount of water by measuring it

- Use a wallpaper paste brush to apply it

- Wash out the paste bucket and brush thoroughly after use.

Filler

Filler is a dry product that again has to be mixed with water in a container. It is then used as a paste which is spread using a filler knife into any surface holes. When dry, filler should be lightly sanded until it is smooth.

Activity: Mixing paste

Prepare your paste bucket by cleaning it before use. Obtain a packet of wallpaper paste and read the instructions on the packet. Measure out the amount of water you need using a measuring jug. Add the required amount of paste and stir until you have the right consistency.

Functional skills

Measuring quantities for mixing paste will help develop your skills in **mathematics**.

✓ Check

- Wallpaper paste has to be mixed correctly, following the instructions

- A variety of different wallpapers are available

L03 Personal Protective Equipment for decorating

Some Personal Protective Equipment (PPE) you will need to use are listed below.

Safety boots

Safety boots have to be worn to prevent any injury to the toes, ankles or base of the foot. Decorators tend to be the last trades that are in a building, completing the decorating. The risk of any objects falling onto feet or trip hazards is therefore reduced. Always wear safety footwear though.

A variety of different safety footwear is available that is comfortable to wear indoors when you are decorating. The shoes below may look like trainers but they are actually safety boots!

Bib and brace overalls

The overalls in the picture are typical of what a decorator would wear. The overalls allow their hands and arms to be flexible when working with a brush and wall coverings, so the worker can move about more easily. They have pockets for brushes, rulers and tapes and a front pocket on the chest for other items.

> ### ✳ Key term
>
> **Barrier**
> A cream that prevents anything getting past and entering your skin.

Hand barrier cream

Because a painter works with many different products that often contain chemicals it is a good idea to use a barrier cream. This is rubbed into your hands before you start work and acts as a 'barrier' preventing any products from coming into contact with your skin.

Dust masks

These are sometimes required when you are undertaking decorating tasks such as:

- Sanding down old plaster

- Sanding down woodwork

- Using chemicals.

> **! Remember**
>
> Gloves should always be used if necessary.

> **! Remember**
>
> You should always check that you have the correct mask for the work you are doing.

Safety hat

Occasionally, when there is a risk of an object falling onto you, or of you hitting your head on an obstruction, the wearing of a hard hat is a requirement.

> **Functional skills**
>
> Reading the instructions on the use of PPE will help you with your skills in English.

Activity: PPE

You have been asked to wallpaper a ceiling. What would be the appropriate PPE to wear?

Check

- PPE should always be used to prevent harm from hazards

L04 Safe working practices

Following advice and guidance

From time to time your tutor will call all of you in to:

- Demonstrate a particular technique

- Give instructions on the next part of your practical

- Advise on any mistakes that you are making.

You will need to take this advice and guidance on board and act upon it, asking any further questions if you do not understand what is required.

Safe maintenance, use and storage of tools and equipment

You will need to make sure that the tools that you are using are:

- Not damaged

- Stored correctly

- Maintained correctly.

Any defective tool will need replacing or maintaining so that it is in a good working order.

Safe working practices

When wallpapering you will need to follow safe working practices:

- Read the instructions on the wallpaper to see what adhesive is required and how long it will need **soaking**

- Check the instructions regarding PPE for its use

- Mix the wallpaper paste correctly into a smooth consistency

- Set up the **pasting table** so that it is level

- Work out how many widths of paper you want and where they will start and finish

- Strike a vertical line down the wall as a plumb guide

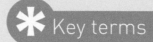
Key terms

Soaking
The amount of time you leave the paste to soak into the wallpaper.

Pasting table
A table used to spread wallpaper out while you apply the paste.

- Measure the wall for the first length of paper, plus 50 mm at the top and bottom for trimming

- Paste the wallpaper and allow to soak

- Hang the wallpaper using stepladders

- Smooth out the wallpaper using a brush

- Trim the top and ends

- Repeat the process until the wall is finished

- Clean the papering table and fold up

- Wash out the paste mixing bucket

- Clean up any off cuts and dispose of safely.

Activity: Wallpaper hanging

Your tutor will give you an area of wall that needs to be wallpapered. Carry out these tasks, following safe working practices:

- Calculate how many widths of wallpaper you require
- Calculate the total length you require, allowing for any pattern repeat
- Cut the wallpaper to length
- Set out a plumb line on the wall
- Follow the instructions and paste the wallpaper
- Hang each width and smooth out, matching the pattern
- Complete the trimming of each piece
- Clean down and tidy your work area
- Ask the tutor to assess your work.

! Remember

Always read the label with the wallpaper to see what the instructions are for applying the paste and hanging the paper.

Functional skills

Measuring the wallpaper, and working out how much you need, will help develop your skills in **mathematics**.

✓ Check

- Hanging wallpaper requires many skills and attention to detail

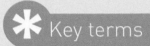

L05 L06 Working responsibly with others

Behaviour

Decorating will often involve working off ladders and platforms. These must be:

- Checked that they are safe

- Used correctly

- Stored correctly

- Have any damage reported.

You must always use the **training**, any information and instructions that you have been given, to ensure that no hazards are created and accidents result.

Hazards

When you work with others, always ensure that:

- You know of any hazards

- You work with your partner safely

- You both know what you have to do.

When you work with other people in a team you have a "duty of care" to take responsibility for yourself and their safety. This is a legal requirement under the **HASAWA** (see *Unit 10 Health and safety and welfare in construction*).

Key terms

Training
Information and instructions given in a formal way.

HASAWA
Health and Safety at Work Act 1974.

 Case study:

Risk assessment

Fred spent most of his day sanding down a wall. He didn't take any time to assess the material on the wall or carry out any risk assessment. By the end of the day his colleague was having problems, coughing severely and having trouble breathing.

- What should Fred have done to prevent this situation?

Working responsibly as part of a team

When you first start working on a site you may have to go through an induction. This is done so that, when working in a team, you know:

- What the fire alarm sounds like

- The evacuation procedure in the event of a fire

- Where the site facilities are

- Where waste materials need safely disposing

- Where any local hazards are.

You will need to make sure that any new decorators that come onto the site are put through the induction so the information is communicated to all in the team.

Responsible team working involves:

- Listening to the supervisor who is in charge

- Acting on the instructions and guidance that they give you

- Reporting any hazards

- Helping when required.

Remember

Always report something that does not look safe. It is better to be safe than sorry.

Functional skills

Communication by talking and listening to a supervisor will help you with your functional skills in **English**.

Activity: Communication skills

A new worker has started in the team doing wallpapering. This person is deaf. How will you communicate?

Check

- Team working involves communicating with all of the team

- Act on the information, instructions and training you have been given

Assessment overview

While working through this unit, you will have prepared for completing the following tasks:

○	1.1	List and describe appropriate hand tools to be used in the decorating process	Pages 100–101
○	2.1	List and describe appropriate materials to be used in the decorating process	Pages 102–103
○	3.1	List and describe appropriate Personal Protective Equipment (PPE) to be used when decorating	Pages 104–105
○	4.1	Select and use wallpaper paste safely to fix wallpaper to a wall of 3 m²	Pages 106–107
○	5.1	Maintain a clean and tidy work environment	Page 108
○	5.2	Work responsibly in the workshop	Page 108
○	6.1	Follow instructions when working with others	Page 109
○	6.2	Communicate appropriately with others	Page 109

Assignment tips

- Keep your work area clean and tidy.

- Keep the wallpaper paste brush on its clip so it does not fall into the paste bucket.

- Keep the pasting table clean and free from dirt.

- Good preparation of a wall before papering will produce a quality result.

- Make sure you use a plumb bob to set out a vertical line to work to.

DEVELOPING PLUMBING SKILLS

For hundreds of years we have required water to be supplied to our homes for drinking, cleaning and washing. It was originally supplied using lead pipes, but today it is supplied through plastic and then copper pipework. Water in a home is heated to provide a supply of hot water for bathing, washing and cleaning. The water supply has to be moved around a home to feed the heating system, drinking water, washing machines, showers and wash hand basins.

In this unit you will learn:

- About the hand tools used in basic plumbing processes

- About the materials and components used in basic plumbing processes

- About the Personal Protective Equipment (PPE) used in basic plumbing processes

- To apply safe working practices to perform plumbing operations

- To work responsibly with others

- To ask for and respond to guidance when working as part of a team

Why is it even more important to follow safe working practices when doing some particular plumbing jobs?

L01 Plumbing tools & equipment

There are many hand tools that may be used during a plumbing task. Here is a selection that are commonly used:

✳ Key term

Blow torch
A flammable gas bottle connected to a jet which gives a directed flame.

	A hacksaw is used to rough cut across pipes.
	A wheel cutter is a tool that is turned around a pipe and cuts a neat line around it.
	Files are used to take off any rough edges to cut pipes.
	Wire wool is used to clean a pipe prior to soldering.
	A wrench is used to tighten any plumbing fittings.
	Grips are used to grip onto fittings to hold them.

	Spanners are used for various plumbing items where nuts need to be tightened.
	A bending spring is slid inside a pipe so it does not crush it when the pipe is bent.
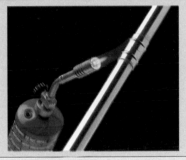	A blow torch is used to heat a joint and melt the solder.

 Remember

Soldering is a hot process and great care should be taken.

As you can see, there is a large variety of plumbing tools that are used in plumbing operations. There are many different fittings that are screwed and tightened together using hexagonal nuts, for example sink taps. Wastes to sinks are normally in plastic and are push fit, so no tools are required.

Activity: Tool identification

Answer the following questions on hand tools for plumbing:

- If a cut pipe has some sharp edges on it and will not fit into a fitting, what tool would you use to smooth these down?

- What is the difference between a wrench, a spanner and grips?

 ## Check

- A number of different tools are used in plumbing, for many different tasks

L02 Materials for plumbing

Some of the materials and components used in basic plumbing processes are listed below and opposite.

Copper pipe

Copper pipe is produced in standard lengths and in different diameters. Common sizes are:

- 15 mm which are used for the branches and water supply

- 22 mm which are used for the hot water to the baths and central heating main runs.

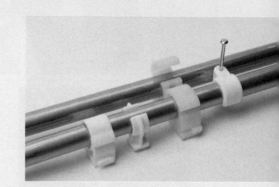

Copper can be easily bent, soldered and connected using brass fittings.

PVC tubing

PVC plastic tubing can be used instead of copper pipework to deliver water around a home.

Plastic tubing has many advantages over traditional copper pipework in that:

- It is flexible

- It does not need soldering

- Installation is faster.

Jointing paste

This is a paste that is made to remain flexible. It is used on screw type fittings to help form a seal which can be taken apart for maintenance if required.

Flux

Flux is used when you are soldering joints, to:

- Provide a cleaning agent for the joint, to remove grease

- Help the solder to flow and bond to the copper pipework.

It is painted onto both pieces that are going to be soldered, with a small brush.

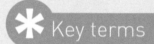
✳ Key terms

Copper
A soft bendable metal that can easily be jointed.

Central heating
The boiler, radiators and pipework systems that heat a home.

Capillary
The action of a liquid to be drawn into a small gap by its surface tension.

Olive
A small copper ring that fits between the pipe and fitting. When tightened, a seal is formed.

Components

A **capillary** joint is a fitting that slots over the copper pipe. It leaves a small gap that is filled by the molten solder by capillary action. They are available with the solder already inside them which melts and is drawn out of the joint by the same action.

A compression joint relies on an **olive** which is compressed around the copper pipe, forming a seal as the nut is tightened. Using a compression fitting means you can easily take things apart to maintain them.

Fitting types

Two common types of fitting are:

- A T-junction fitting, which is used to form a branch off a main pipe run

- A bend fitting, which is a 90 degree fitting and is the most common bend used.

Activity: Compression fittings

- Washing machines require a compression fitting to connect them to the water supply. Find a picture of this fitting.

 Functional skills

Taking a look through plumbing suppliers' catalogues will help with your reading and improve your skills in **English**.

Check

- There are many different methods of joining used in plumbing

L03 Personal Protective Equipment for plumbing

When carrying out basic plumbing processes make sure you are using the correct Personal Protective Equipment (PPE).

Safety gloves

The soldering of copper pipework involves heating the pipe with a flame. This makes the copper conduct heat and it can be hot to the touch. Often gloves will need to be worn when handling such items. Gloves would also need to be worn where there is risk of injury to the hands – for example when using wrenches, grips or spanners.

Goggles

Eye protection will need to be worn if you are using equipment that would cause particles to fly off towards your face.

Safety boots

It is good practice to ensure that safety boots are worn as there are many hazards on construction sites.

Forming capillary joints

The process to solder a copper joint is shown in the flow diagram (right).

Key terms

Soldering
Using a blow torch and melting solder to seal joints in pipes.

Joint
The finished sealed fitting.

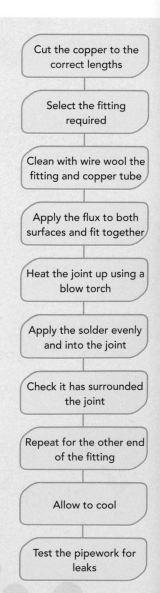

Cut the copper to the correct lengths

Select the fitting required

Clean with wire wool the fitting and copper tube

Apply the flux to both surfaces and fit together

Heat the joint up using a blow torch

Apply the solder evenly and into the joint

Check it has surrounded the joint

Repeat for the other end of the fitting

Allow to cool

Test the pipework for leaks

Forming compression joints

The process in the formation of a compression joint is shown in the flow diagram (right).

Activity: Produce a capillary joint

- Produce a capillary joint using 15 mm diameter pipe.

Bending copper pipe

This is the easiest method of forming a bend in copper pipework. It uses a curve and a guide to bend the pipe to a radius. It saves time in forming soldered joints.

Connecting taps and traps

Taps are connected by pushing the tap through the hole in the sink. A plastic top hat washer normally secures the tap to the sink. A tap connector fitting is used which screws onto the bottom of the tap and the pipe serving it.

Activity: Identifying taps

Take a look at your kitchen or bathroom sink. Examine the taps and the waste system.

- What type of trap have you got installed?

- How are the taps attached to the sink?

Check

- Good preparation before soldering will ensure a watertight joint

---

Flow diagram (right side):

- Select the correct compression fitting for the diameter of the pipe
- Cut the pipe to the correct length
- Take the compression fitting apart
- Slide the locking nut over the pipe
- Slide the brass olive over the pipe
- Fasten the main body of the fitting to the nut
- Repeat for the other end of the fitting
- Hold the main body with grips and tighten both of the nuts using a wrench or adjustable spanner
- Check for leaks

L04 Performing plumbing operations safely

Following advice and guidance

From time to time your tutor will call all of you in to:

- Demonstrate a particular technique

- Give instructions on the next part of your practical

- Advise on any mistakes that you are making.

You will need to take this advice and guidance on board and act upon it, asking any further questions if you do not understand what is required.

Safe maintenance, use and storage of tools and equipment

You will need to make sure that the tools that you are using are:

- Sharp

- Not damaged

- Stored correctly

- Safe.

Any defective tool, especially when dealing with gas bottles, will need replacing or maintaining so that it is in a good working order.

Copper pipe rig with capillary joints

This is a pipe **rig** that has a number of different capillary fittings and some pipe bends within it. Your tutor will set out the sizes for you to have a go at working on this in the practical sessions.

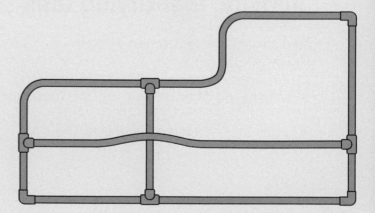

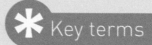
Key terms

Maintenance
Making sure that a tool is in good order and not broken.

Rig
A small-scale pipework assessment piece.

◎ ## Activity: Capillary practical task

Copy down the drawing for the pipe rig with the dimensions that you will need and produce the pipe rig fully soldered.

Compression joints

This is a pipe rig that has a number of different compression joints, and T joints, elbow fittings and some capillary pipe 90° bends within it. Your tutor will set out the sizes for you to have a go at working on this in the practical sessions.

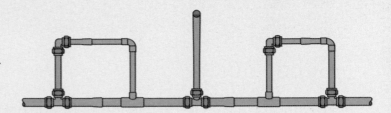

Activity: Compression practical task

Copy down the drawing for the pipe rig with the dimensions that you will need and produce the pipe rig fully soldered.

Connection of hot and cold water taps and tubular swivel trap to a sink

As part of the assessment for this unit, you will have to fit hot and cold water taps to a sink and connect up its waste pipe system, then test it for leakage. To do this you will need to follow the following sequence safely:

- Make sure that the water supply is off

- Fit the sink unit to the base unit

- Fit the hot and cold taps to the sink unit, using top hat washers

- Prepare the 15 mm waste connections by soldering onto the pipework the tap connection pipework you have made

- Fit the hot and cold water tap connectors and tighten using a tap spanner

- Fix the waste P trap by screwing to the waste outlet on the sink

- Measure and cut the correct length of waste pipe and connect the trap to the drain

- Fully test the sink unit to make sure it works correctly.

Check

- Keeping your work clean, neat and tidy will produce a quality outcome

L05 L06 Working responsibly with others

Behaviour

When you are working within the workshop always make sure that you act sensibly and show the correct **behaviour**. For example, a special area should be made available for soldering joints, that:

- Is a safe fire proof location

- Is free from obstructions

- Has a cooling bay for your work

- Contains the wire wool, flux, solder and blow torch.

When you work with others always ensure that:

- You know of any hazards

- You work with your partner safely if a rig needs holding

- You both know what you have to do.

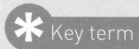

Key term

Behaviour
The way in which you act – your conduct or manner.

Activity: Working responsibly

Think about the following scenario and identify what could be the possible cause of this injury and how it could have been avoided.

James had just finished soldering his 90° capillary joint. He had put down the blow torch and picked up the work straight away, burning the fingers on his right hand.

Strengths and skills

Working within a team often involves looking at your own and others' strengths and skills. Within plumbing this could be:

- The ability to solder correctly

- Measuring dimensions accurately

- Pipe cutting correctly.

Remember

Don't be afraid to ask a colleague or your tutor for help, as you will learn these new skills by example and demonstration.

Working as part of a team

Behaviour

A team needs a good leader. This person should:

- Be knowledgeable about the work

- Be able to get on with other team members

- Be friendly and approachable

- Understand other team members' views

- Be able to give instructions.

Key term

Enthusiastic
Showing an interest in the
work that you are producing.

Attitude

You must maintain a sensible attitude when working as part of a team. This sets a good example for the rest of the team to follow and prevents any disturbance or upset from clients, colleagues or members of the public when working on a site.

Enthusiasm

This is an essential skill that you will need when working with others. People who are continually late and leave early, or have a poor sickness record, usually are not bothered about the job and don't appear **enthusiastic**. They may not be employed for very long.

Check

- Act on the information, instructions and training you have been given

- Team working involves using other people's strengths and skills

Assessment overview

While working through this unit, you will have prepared for completing the following tasks:

○	1.1	List and describe appropriate hand tools to be used in basic plumbing processes	Pages 112–113
○	2.1	List and describe appropriate materials to be used in basic plumbing processes	Pages 114–115
○	2.2	List and describe appropriate components to be used in basic plumbing processes	Pages 114–115
○	3.1	List and describe appropriate PPE to be used in basic plumbing processes	Pages 116–117
○	4.1	Select and use hand tools safely to connect copper tubes	Pages 118–119
○	4.2	Select and use hand tools safely to install a functioning sink	Pages 118–119
○	5.1	Maintain a clean and tidy work environment	Page 120
○	5.2	Work responsibly in the workshop	Page 120
○	6.1	Follow instructions when working with others	Page 121
○	6.2	Communicate appropriately with others	Page 121

edexcel :::

Assignment tips

- Keep your work area clean and tidy.

- Always use the hot work area to solder.

- Make sure that you clean the copper pipe and fitting well.

- Always use a flux inside the fitting before you assemble it for soldering.

- Don't pick up recently soldered work – it's hot.

- Ensure compression fittings are tightened sufficiently.

DEVELOPING ELECTRICAL INSTALLATION SKILLS

Electricity was first introduced into our homes just over 100 years ago with the development of the light bulb. Modern homes now have electrical power and lighting at the flick of a switch to provide energy.

In this unit you will learn:

- About the hand tools used in basic electrical installation processes

- About the materials and components used in basic electrical installation processes

- About the Personal Protective Equipment (PPE) used in basic electrical installation processes

- To apply safe working practices to perform electrical installation operations

- To work responsibly with others

- To ask for and respond to guidance when working as part of a team.

Why does an electrician need a tool belt?

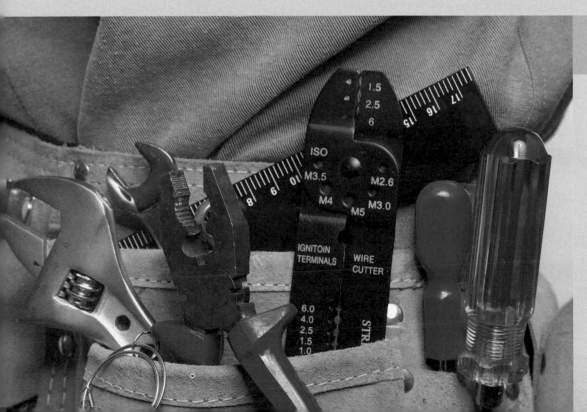

123

(L01) Electrical installation tools

The hand tools used in basic electrical installation processes are shown below and opposite.

✳ Key term

Multimeter
An electrical instrument that can read different voltages and amps.

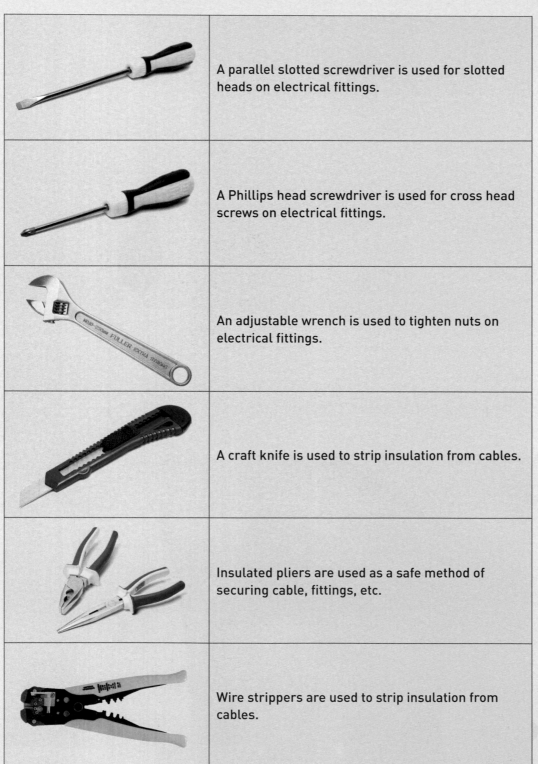

	A parallel slotted screwdriver is used for slotted heads on electrical fittings.
	A Phillips head screwdriver is used for cross head screws on electrical fittings.
	An adjustable wrench is used to tighten nuts on electrical fittings.
	A craft knife is used to strip insulation from cables.
	Insulated pliers are used as a safe method of securing cable, fittings, etc.
	Wire strippers are used to strip insulation from cables.

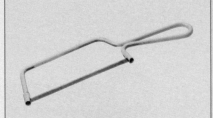	Junior hacksaws are used for cutting larger cables.
	Digital multimeters are used for electrical testing.

Always use electrical tools that are insulated. In order to carry your electrical tools safely while working you will need a tool belt similar to this one.

This holds the tools safely and keeps them from getting damaged. It has pockets to carry additional tools.

Activity: Tool identification

Identify the tools that you would need for each of these electrical installation tasks:

- Wiring a plug

- Cutting mains cable to length.

✓ Check

- A number of different tools are used in electrical installation work

- Tools must be insulated for safety

L02 Electrical installation materials

Electrical cabling

Electrical cabling has three wires:

- A live wire which is coloured brown

- A neutral wire which is coloured blue

- An **earth** wire which is not coloured and requires a sleeve fitting or is yellow and green.

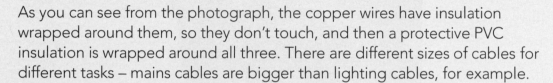

As you can see from the photograph, the copper wires have insulation wrapped around them, so they don't touch, and then a protective PVC insulation is wrapped around all three. There are different sizes of cables for different tasks – mains cables are bigger than lighting cables, for example.

Components used in basic electrical installation processes are listed below and opposite.

Sockets

13A sockets are used to provide outlets for you to plug your electrical equipment into. The plug that is on the end of the cable has a **fuse** within it. Sockets are placed around a home on a wiring system known as a ring main, as it loops in and out of each socket supplying power.

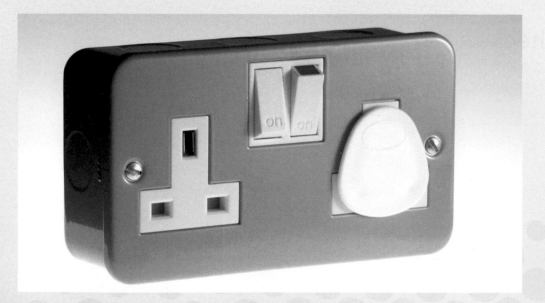

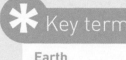
Key term

Earth
A safety feature where any electricity is returned to the ground safely.

Fuse
A device that gets hot if overloaded and melts, breaking a circuit.

Lighting

Lighting is provided in a home by the use of a light fitting and a bulb to produce the light. These are manufactured in many different shapes and sizes, from external lights to internal glass and brass fittings.

Fittings

Light fittings are manufactured to hold a light bulb. Since there are many different light bulbs there are many different light fittings. There is a range of fittings to cover:

- External lighting which must be waterproof
- Internal light fittings for decoration and different types of light
- Fittings for different rooms, for example a shower room.

Light bulbs

Light bulbs are manufactured in a range of wattage. This is the amount of energy the bulb uses and hence how bright it is going to be when you switch it on. There are two types of fitting:

- Bayonet, which uses two pins on the light bulb to secure it
- Screw fittings, which have a thread on the bulb that screws into the fitting.

Fuses

Fuses are placed into plugs and have different ratings. Typically 3A and 13A cartridge fuses are fitted to electrical equipment.

Activity: Check the fuse

Choose a piece of electrical equipment and disconnect it from the electrical supply. Remove the cover to the plug and check that it has the correct fuse fitted.

Check

- Electrical wiring has to be done safely using the right materials for the task

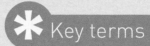
L03 Personal Protective Equipment

When carrying out basic electrical installation processes make sure you are using the correct Personal Protective Equipment (PPE).

Safety gloves

Gloves should be worn to:

- Cut metal conduit with a **hacksaw**
- Cut large cables with cutters
- Handle any sharp materials.

Goggles

Eye protection will need to be worn if you are using equipment that would cause particles to fly off towards your face.

Safety boots

The soles of safety boots are made from an insulating material. This stops an electric current passing through your body to earth and keeps you safe.

Know what PPE you need for the tasks you will be doing.

Installing a circuit

Isolate power supply

This is a very important operation. You must ensure that the power is switched off to prevent any electrical **shock** while you are work.

Make sure that the power supply is set to off and locked off by your tutor so it cannot be switched on without their intervention.

Mark out circuit

You may have been given a drawing of the circuit. The dimensions on it will need transferring to the board you are going to install your circuit on.

✱ Key terms

Hacksaw
A frame with a handle that holds a toothed blade used to cut metals.

Shock
A discharge of electricity through your body to the earth causing muscle contraction.

! Remember

Always use the correct PPE when carrying out electrical installation tasks.

Position and fix components

The drawing will show you where any components, such as sockets and switches, will need installing. Mark out their positions.

Measure cables to length

Cables are kept on reels. You will need to correctly work out how much you need from the drawing, then measure and cut to length using pliers the right amount of cable.

Strip cables and wires

Using the cable stripping tool, take off the required amount of insulation to the cables so they can be fitted into the components. Cut the required amount of earth sleeving and push this over the earth wire.

Fix wires to components

To do this you will need a good slotted electrical screwdriver. You will use this to open and tighten the terminals on the electrical switched sockets. Take care that you use the right size as the threads and heads of the screws can become easily damaged.

Check connections and test system

This must be done before you switch on the power and make your board live.

Activity: Connecting a socket

Produce a connection to a 13 Amp double socket, using 2.5 mm twin and earth.

Activity: Wiring up a cable

Wire up the cable using a suitable plug fitting a 5 Amp fuse.

Check

- When cutting cables to the correct length, it is good practice to make sure all the exposed copper is shielded

L04 Performing safe electrical operations

Follow advice and guidance given

Electricity is dangerous and can kill. An electrical cable carries 13 Amps of power while 1/2000th of an Amp can stop your heart. You will need to listen carefully to instructions given by your tutor. These will include:

- How to isolate the electrical supply

- Ensuring correct wiring

- Ensuring all socket faces are on and all wires in conduits

- Testing of the circuit.

Safe maintenance and storage of tools and equipment

You will need to make sure that the tools you are using:

- Have electrical insulation on them

- Work correctly, for example a multimeter

- Are sharp for cutting wires and insulation.

Any broken tools will need replacing or maintaining so that they are in good working order.

Now that you have thought about how to work safely you will be asked to build something like the circuit on the right. This is the way in which lights are wired in parallel. So when the circuit is switched on they all come on. This is the simplest way to wire up several lights that need to be switched on together.

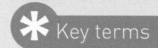

Key terms

Operations
A task, for example isolating a circuit.

Basic ring main

You will be assessed on wiring up a basic ring main using four 13 Amp fused and earthed sockets. Take care to ensure that you get the wires in the right order into each socket.

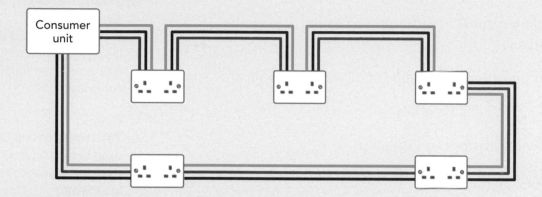

Activity: Produce a lighting rig

Produce a lighting rig with four lights wired in parallel on the rig board. Test this circuit when complete.

! Remember

You will need to use a multimeter to check that the circuit is complete before making the circuit live.

Activity: Produce a ring main

Produce a ring main with four sockets wired on it. Test this mains circuit when complete.

✓ Check

- Plan your work, cut cables carefully and follow the wiring diagram to produce the electrical installation

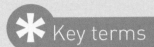

L05 L06 Working responsibly with others

Maintain a clean and tidy work environment

Good housekeeping is essential in order to maintain a safe working environment. This means:

- Not leaving any sharp objects on the floor or working surface

- Removing all packaging which is a fire risk

- Cleaning away all wire offcuts and insulation

- Placing all tools and equipment into tool bags and storing after use

- Emptying rubbish bins and disposing of waste safely.

Working responsibly

Working with electricity is dangerous. You must not:

- Switch on any live supply to your board unless a tutor is present

- Leave any **exposed** copper wiring

- Cut a live wire

- **Plug** anything into a socket without your tutor present.

> **Key terms**
>
> **Exposed**
> Wiring that has had the insulation removed from it to expose the copper core wire.
>
> **Plug**
> The connector that is used to connect a wire from the appliance to the mains.

Activity: Acceptable behaviour

Simon is wiring his four sockets onto the board using a ring main. The tutor has isolated the electrical supply. Simon's working colleague asks the tutor to make the board live.

- What could be potentially dangerous here?

! Remember

Always report or isolate exposed wiring to remove the hazard.

Respond to guidance

You must always use the training, and any information and instructions that you have been given, to ensure that no hazards are created and accidents result. When you work with others in performing electrical installation always make sure that:

- Your colleague knows that the electrical supply is switched off
- You work with your partner safely
- You both know what you have to do
- You test the circuit before making it live.

If you are in any doubt ask your tutor for help, information and guidance on the electrical installation task you are performing.

Communication

Working with electricity can be hazardous. To avoid potential hazards 110 volts is used for the temporary work on site. This is done to:

- Reduce the risk of electric shock
- Reduce the Amps that are present in the 110 volt cable.

Electrical cabling is colour coded on a construction site so you know what the voltage is in the cable:

- Yellow is 110 volts
- Blue is 240 volts
- Red is 415 volts.

Each of these cannot be plugged into each other as the plugs are different. This prevents a serious accident and electrical fire.

Functional skills

Communication by reporting hazards will help improve your skills in speaking **English**.

Activity: Switch off

How will you ensure that all in a team know that the power has been switched off?

Check

- Act on the information, instructions and training you have been given
- Team working involves looking after other people's safety and not just your own

Assessment overview

While working through this unit, you will have prepared for completing the following tasks:

◯	1.1	List and describe appropriate hand tools to be used in basic electrical installation processes	Pages 124–125
◯	2.1	List and describe appropriate materials and components to be used in basic electrical installation processes	Pages 126–127
◯	3.1	List and describe appropriate PPE to be used in basic electrical installation processes	Pages 128–129
◯	4.1	Select and use hand tools safely to perform basic electrical installation operations	Pages 130–131
◯	5.1	Maintain a clean and tidy work environment	Pages 132–133
◯	5.2	Work responsibly in the workshop	Pages 132–133
◯	6.1	Follow instructions when working with others	Pages 132–133
◯	6.2	Communicate appropriately with others	Pages 132–133

Assignment tips

- Keep your work area clean and tidy.

- Cut cables and insulation correctly.

- Always isolate the power supply.

- Listen to instructions given by your tutor.

- Follow wiring diagrams closely.

- Always test a circuit before switching on the power.

DEVELOPING BUILDING MAINTENANCE SKILLS

Maintenance is how we keep our homes in good condition. People in the UK spend millions of pounds renovating and maintaining homes. Maintenance is essential in order to keep the weather from damaging the structure through the actions of wind, rain and frost. Building maintenance and improvements to our homes often involve a number of different trades. Painting, decorating, plumbing and electrical work can all be involved, for example from repainting the outside of a house to fitting a new light.

In this unit you will learn:

- About the hand tools used in building maintenance processes

- About the materials used in building maintenance processes

- About the Personal Protective Equipment (PPE) used in building maintenance processes

- To apply safe working practices to perform building maintenance tasks

- To work responsibly with others

- To ask for and respond to guidance when working as part of a team

How do simple things like making sure a gutter is clear help maintain a building?

LO1 Building maintenance tools & equipment

The following table illustrates some of the common hand tools that you would use for home improvements and maintenance around a house.

	A pointing trowel is used to repair joints in brickwork.
	Step ladders are used for access during maintenance jobs.
	A builder's bucket is useful for mixing mortar and rubbish removal.
	A screwdriver set is used for many home improvement and maintenance tasks.
	A tape measure is used for measuring items.
	A cordless battery drill is an essential piece of equipment.
	A drill bit set is used with the cordless drill.

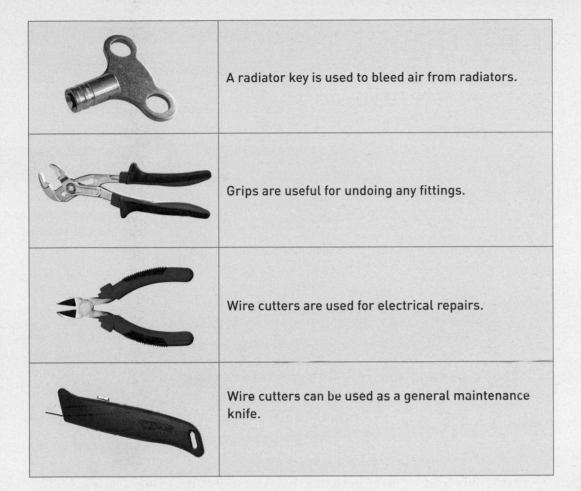

	A radiator key is used to bleed air from radiators.
	Grips are useful for undoing any fittings.
	Wire cutters are used for electrical repairs.
	Wire cutters can be used as a general maintenance knife.

Activity: Which tool?

Identify the tools that you would need for each of the following building maintenance tasks:

- Repair a sticking wooden door
- Repair a sink waste that is blocked
- Repair a leaking washing machine pipe joint.

What tools would you need for the following home improvement tasks:

- Hanging a picture
- Painting a window frame
- Changing a door lock.

To get to the tap washer you will need to take off the tap knob and unscrew the washer housing.

✔ Check

- There is not just one tool kit for home and building maintenance – you will need to collect tools as you progress

L02 Carpentry maintenance & improvement

Fixings

There are a number of different **fixings** that are used for home improvements.

Screws and plugs are the most common for use on solid walls. Holes have to be drilled using a masonry drill bit.

Hollow partition walls (which are made using studs and covered with boards) will need special fixings that grip the back of the plasterboard and tighten onto it.

Anything fixed to timber needs a screw to secure it, for example a hinge, bracket or fixing.

Carpentry

You might need to carry out repairs and fix fittings to internal and external joinery items such as:

- Window frames

- Guttering and downpipes

- Doors

- Shelving

- Putting up pictures

- Kitchen cupboard

- Curtain rails.

✳ Key term

Fixings
Used to fasten and hold one thing to another.

Case study:
Starting your own business

Phillip had trained as a carpenter for three years at his local college and had set up his own business undertaking timber maintenance works. He had in the last month looked at and completed jobs on:

- Roofing work – replacing some roof tiles that had been broken

- Guttering – replacing old timber guttering that had rotted with new uPVC guttering and downpipes

- Broken windows – renewing glass in the window sash

- Sticking doors – easing doors that had swollen with the rain and weather

- Fixing floor joists – for a builder constructing a house at first floor level.

All of this work had been done for different clients in the surrounding area where he lived. Phillip now has so much work on that he has bought a van to transport his materials and tools around safely.

Read through the case study again and then discuss the following questions in small groups:

- What would motivate you to start your own business?

- How would you go about this?

- What extra skills will you need to develop the business?

Check

- Working as a carpenter may mean that you have to undertake jobs that are not specifically timber related

LO2 Plumbing & heating maintenance & improvement

Radiators

Maintenance and improvement works to radiators may include:

- Installing a decorative MDF radiator cover – this is a special casing that fits around the radiator and is painted

- Disconnecting a radiator to paint or paper behind it, then putting it back again.

General plumbing items

Listed in the table are some typical items that you might improve or maintain and what you need to do.

Item	Problem	What you need to do
Seal round a bath or shower	The sealant often becomes discoloured with age	Replace with new sealant
Washers	Washers compress and wear out	Take apart the fitting with the washer in it so it can be renewed to stop leaking
Remove air-locks and balance radiator systems	In some central heating systems air can build up within radiators, causing problems	Air needs to be 'bled' from the radiators by opening the vent and letting it out
Loft insulation	Heat can be lost through the roof	Rolling out mineral wool insulation in the loft traps air which keeps the heat in
Lag pipes and cisterns	Pipework can freeze and burst during the winter	Insulation placed around any cold water pipework can prevent this

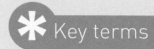

Key terms

MDF
Medium Density Fibreboard.

Radiator
A panel which is hollow, fills with hot water and emits heat.

Activity: Removing a radiator

Look at this description of how to remove a radiator for a decorator to wallpaper behind. Can you find anything wrong?

- Step 1 – Remove air valve

- Step 2 – Bleed radiator from air valve

- Step 3 – Empty radiator

- Step 4 – Disconnect radiator valves from radiator

- Step 5 – Remove wall brackets

- Step 6 – Check for leaks.

Case study:

Burst pipes

It is the middle of winter and a client has called you out to a burst water pipe. They have managed to switch off the water. The pipe is an outside supply to a tap where it is on the outside wall. There has been some flooding outside, as the pipe had run all day until the client arrived home. They are quite annoyed because it has caused damage to their garden and outbuildings.

- What tools and equipment will you need to carry out this work?

- Are there any precautions you will need to take?

Check

- Plumbing maintenance involves any work that is connected with water supply and disposal

L02 Electrical maintenance & improvement

Wire plugs

Most new equipment has a **plug** already on it, but now and then you may need to renew a damaged plug. To do this you will need to follow the following sequence safely.

- Remove the old plug by unscrewing the cover and releasing the wire terminals and cable grip

- Remove the new plug cover and look closely inside the new plug to see if the existing wires need extending or shortening

- Remove the cable grip if this has screws attached

- Fit the wires into the pins and tighten the screws

- Replace the cable grip and tighten if it has screws

- Refit the plug cover

- Test the electrical item is working.

Changing fuses

A **fuse** is a safety device. If it has 'blown' then this has happened for a reason and you should follow the following advice:

- Switch off and disconnect the power supply

- Find out if there is any damage to the equipment or cable

- Repair this

- Remove the plug cover or fuse carrier

- Check the fuse rating

- Replace the fuse

- Test the item is now working.

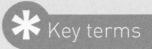

＊ Key terms

Plug
A device to connect equipment to the mains.

Fuse
A device that gets hot if overloaded and melts, breaking a circuit.

! Remember

Always disconnect the electrical item you are working on from the power supply.

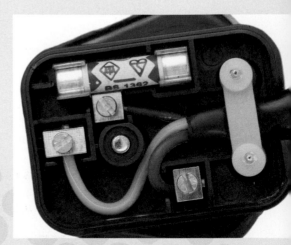

Putting in a simple light fitting

Often light fittings are improved and upgraded. To do this you will need to remove the old fitting first. The following is a list of operations you will need to follow:

- Switch off the power supply
- Remove the existing light shade and bulb
- Unscrew the cover which is hiding the electrics on the ceiling
- Disconnect the electrical wires – live, neutral and earth
- Remove the complete fitting
- Read the new fitting's instructions
- Fit the new rose to the ceiling, passing the wires into it
- Connect the existing wires to the terminal block
- Fix the light, fitting live and neutral, and screw the cover in place
- Fit the shade and bulb
- Test.

Activity: Which tool?

Take a close look at a light fitting. After turning off the electrical supply, carry out the following:

- Remove the bulb
- Remove the lamp shade
- Clean the shade
- Refix the shade and bulb
- Test the light.

What should you always do before changing a light?

Check

- Working with electricity is dangerous – always be cautious
- Make sure you test after you complete electrical maintenance jobs

L02 Brickwork, tiling & plasterwork maintenance & impovement

Brickwork repairs

The type of repair that may be required for brickwork includes:

- **Re-pointing** of brickwork where the mortar in the joints has fallen or been worn out.

The diagram on the right shows you what to do to make this repair.

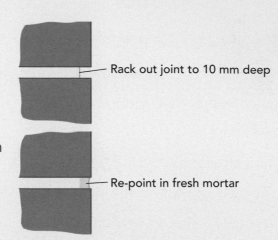

Rack out joint to 10 mm deep

Re-point in fresh mortar

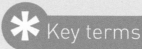

Key terms

Re-pointing
Removing old mortar and filling with new.

PVA
Polyvinyl acetate glue.

Activity: Re-pointing brickwork

Your tutor will identify a brick panel that has been built using a lime mortar. Rack out a joint to a depth of 10 mm and re-point this joint to an acceptable standard.

Plaster repairs

Older properties where plaster has aged may need some patching undertaking where damage may have occurred. To undertake a plaster repair you will need to:

- Remove old plaster back to a sound edge

- Prepare the background with **PVA** bonding agent

- Mix the first coat of background plaster

- Apply the first coat and leave to set

- Use a tool to scratch the surface so the next layer sticks properly

- Mix and apply the finishing coat and trowel smooth

- Leave to dry.

Tiling

Tiling work is required in bathrooms, showers and sink splash backs. It provides a clean, waterproof and washable surface and is available in a range of different shaped and patterned tiles. To apply tiles you will need to do the following:

- Gather the tiles and adhesive that you need

- Mark out the wall for your tiles, making sure you have a vertical and horizontal line to follow

- Apply the adhesive over small areas at a time

- Fix the tiles to the adhesive and use tile spacers to even them out

- Allow the adhesive to dry

- Mix the required amount of tile grout and press into the space between the tiles

- Allow to set and rub off the excess, leaving a neat joint.

Activity: Tile fixing

- Prepare a wall area and fix four tiles to the wall above a sink to form a splash back.

✓ Check

- When applying adhesive do small areas at a time

- Maintenance and repair work requires good preparation for it to be effective

L03 Personal Protective Equipment

When carrying out building maintenance processes make sure you are using the correct Personal Protective Equipment (PPE).

Hard hat

Domestic repairs at home might not need a hard hat. But it is good to start to learn to wear one on all jobs where there could be a risk of any materials falling on your head or of you banging your head against an obstruction.

Eye protection

Eye protection must be worn where you are undertaking tasks where dust or debris could enter your eyes and cause an injury. Safety glasses or safety **goggles** are available, depending upon the task you are doing and the level of protection that is required.

High-visibility jacket

A **high-visibility** jacket is a yellow (or orange) jacket that is worn over your clothes or overalls. It is 'high' in colour, so people who are driving plant and equipment can see you and avoid you. They are especially good in low light conditions and often have reflective strips on them for use in the dark. They are also available as a bib for more comfortable wearing in hot weather.

Safety boots

Safety boots need to be worn on sites and at home to prevent any injury to your toes when handling materials. There is also a variety of other safety footwear, such as shoes and trainers available.

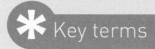
* Key terms

Goggles
PPE that covers the whole of the eyes with a flexible plastic shield.

High-visibility
Uses a contrast colour, usually yellow or orange, that is visible to the eye.

Case study:
Wearing PPE

Tom had just started work on a construction site as the assistant site manager. He had been issued with the following PPE:

- Safety boots

- Hard hat

- Site coat

- Gloves

- High-visibility vests.

Tom wears all of this on the first day. After about a month Tom started to leave his high-visibility vest in the cabin as it was hot during the day. He also stopped wearing his hat inside the building.

Carefully read over the above case study. How you would handle a worker on site refusing to wear their hard hat?

Check

- PPE is important and must be worn to set a good example

- This includes all the PPE you are provided with on site

L04 # Building maintenance tasks in brickwork & carpentry

Brickwork

Here are some typical maintenance repairs to external brickwork in walls:

- Cracks in brickwork – this will involve chopping out the cracked brick and its surrounding mortar and replacing it with a brick to match.

- Damaged pointing and rendering – damaged pointing will need racking or cutting out, then re-pointing. Cracked or damaged render will need removing. The edge is then treated with a PVA bonding agent. Render is then used to finish the area.

- Broken roof tiles – great care must be taken with this as you may be working at height. Scaffolding, or roof ladders, may need to be used to provide a safe means of access and egress.

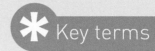

Key terms

Render
A sand and cement mixture that acts as the first coat before plaster.

Access and egress
Getting in and out or up and down.

Activity: What PPE?

What PPE would you need to wear if you were removing and replacing a damaged brick?

Carpentry and joinery

Here are some typical maintenance repairs to carpentry and joinery items.

- Doors and hinges – doors and hinges may require re-securing, fixing or replacing if they are damaged. Doors may need easing if they are sticking or have become warped.

- Windows – broken windows will need replacing if they are cracked, damaged or if the double glazed unit seal has failed.

- Flooring/loose and broken boards – floorboards over time will ease and may need re-fixing. Split or damaged boards will need replacing. Boards may need to be lifted for any electrical or plumbing work.

- Skirting boards – skirting boards do not often require any maintenance work. They may require attention where doors are moved or if damage has occurred during carpeting works.

- Rotten timber – external timbers will suffer due to water rotting the wood. The wood will need cutting out and replacing with new timbers.

Activity: Outside maintenance

- What kind of maintenance work does the outside of a house need?

Check

- Bricks, tiles, cement, doors, hinges and timber need the most maintenance

- Maintenance should be carried out on a regular basis

Building maintenance tasks in
LO4 painting & decorating, plumbing & electrical maintenance

Painting and decorating

Painting must be maintained as the weather will gradually break the surface down. Undercoat and gloss finishes will need to be re-applied every couple of years in order to maintain the level of protection.

Paintwork always requires preparation by sanding down and removing any dust or dirt before starting repainting. Any surface dents should be filled and smoothed.

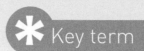
Key term

Components
Parts of something.

Plumbing

Typical maintenance repairs to plumbing items include:

- Air locks in pumps and radiators

- Blockages in drains, basins, toilets and gullies

- Failed **components**, such as washers, ball valves, diaphragms and leaking joints.

Electrical maintenance

Typical maintenance includes:

- Failed fuses will require replacement

- Non-working switches will need replacing

- Broken switches and non-working sockets will need to be replaced

- Broken or cracked plugs will need replacement

- Ceiling roses may become damaged over time

- Radiator valves often stick and may need easing or replacement.

Activity: What needs maintaining?

Look around a room at home or in your school or college. How many painting and decorating, plumbing and electrical items can you see that might need to be maintained?

✓ Check

- There is a wide range of items within a home that will need maintaining

- Always follow safety guidelines when carrying out electrical maintenance tasks

L05 L06 Working responsibly with others

Maintaining a clean and tidy work environment

Good housekeeping is an essential requirement of any working environment. If you keep a work area clear of waste materials, unused tools and trailing electrical leads, it will prevent many small accidents. Keeping this up will make working **colleagues** follow your example.

Working responsibly in a workshop

Workshops can be dangerous places. They are full of equipment and machinery, dust and noise that contain many hazards. When working in this environment you should:

- Work safely

- Have regard to the safety of others around you

- Alert all to any hazard that is harmful.

Behaviour

Building maintenance work will especially mean that you have to work with other people and show the correct **behaviour**. This could involve:

- Supporting the base of a ladder for a colleague

- Passing materials up

- Lifting materials

- Assisting by holding and supporting

- Cutting while your colleague fixes.

You should make sure that both of you can hear each other and can give and receive instructions so that the task you are undertaking is safe. This may involve:

- Having to shout above noisy work

- Asking your colleague to repeat themselves

- Using hand signals

- Double-checking before acting.

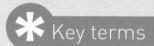

Key terms

Colleagues
The people that you are working with.

Behaviour
The way in which you act – your conduct or manner.

Case study:
Working responsibly

Daniel and Jonathan were working on a roof at first floor level on an extension to a house where two roof tiles had been broken. They had set up some ladders and Daniel was working on removing the tiles at arms length on the top of the ladders. Jonathan then answered his phone and just wandered away talking to their boss.

- What could happen in this situation?

Communication

Maintenance work may involve working in a team with each person having different roles and responsibilities:

- A plumber will do water and heating maintenance
- An electrician will look at electrical faults
- A plasterer may make repairs to finishes
- A painter and decorator will improve and repair decorations.

Multi-skilling is the way forward for home improvements and maintenance. This means you can do more than one job, for example a bricklayer can do some plastering and a carpenter can repair a roof tile and change a fuse or plug.

Functional skills

While working within a team, speaking and listening will help you with your skills in **English**.

Activity: Employment opportunities

Have a look at a large commercial organisation near you. What jobs do they offer there in maintenance?

Check

- Working with others must be taken responsibly
- Team working in maintenance may mean doing more than one type of task

Assessment overview

While working through this unit, you will have prepared for completing the following tasks:

○	1.1	List and describe appropriate hand tools to be used in building maintenance processes	Pages 136–137
○	2.1	List and describe appropriate materials to be used in building maintenance processes	Pages 138–145
○	3.1	List and describe appropriate PPE to be used in building maintenance processes	Pages 146–147
○	4.1	Select and use hand tools safely to perform building maintenance tasks	Pages 148–151
○	5.1	Maintain a clean and tidy work environment	Page 152
○	5.2	Work responsibly in the workshop	Pages 152–153
○	6.1	Follow instructions when working with others	Pages 152–153
○	6.2	Communicate appropriately with others	Page 153

Assignment tips

- Look carefully at what maintenance job has to be done – it may involve several operations and trades.

- Always isolate any electrical supply.

- Find out where the water can be turned off when renewing tap washers.

- When removing a radiator have an old towel ready for spillages.

- Always test an item before making it live.

Key terms

Access and egress – Getting in and out or up and down.

Accidents – An unplanned event resulting in damage or injury.

Barrier – A cream that prevents anything getting past and entering your skin.

Behaviour – The way in which you act – your conduct or manner.

Bevel – A slope or angle.

Blow torch – A flammable gas bottle connected to a jet which gives a directed flame.

Border – A box that surrounds the drawing.

Buttering – Placing mortar on the end of the brick or header.

Capillary – The action of a liquid to be drawn into a small gap by its surface tension.

Central heating – The boiler, radiators and pipework systems that heat a home.

Coat – One application of a paint finish which is left to dry.

Colleagues – The people that you are working with.

Communication – Talking and listening.

Components – Parts of something.

Contractor – A company that has been given a contract to do part or all of a construction project.

Copper – A soft bendable metal that can easily be jointed.

Cordless – A power tool that does not plug into an electrical supply.

COSHH – The Control of Substances Hazardous to Health regulations.

Cutting in – Finishing correctly where paint touches another of a different colour.

Dry bonding – Laying bricks in position with no mortar to check the length of the wall.

Earth – A safety feature where any electricity is returned to the ground safely.

Egress – See *Access and egress*.

Enthusiastic – Showing an interest in the work that you are producing.

Environment – The surrounding area you are working in.

Exposed – Wiring that has had the insulation removed from it to expose the copper core wire.

Extinguish – To put out a fire or flame.

Fixings – Used to fasten and hold one thing to another.

Fixtures and fittings – For example, a lamp, light or furniture.

Forklift trucks – A mechanical means of lifting using two forks.

Fuse – A device that gets hot if overloaded and melts, breaking a circuit.

Gauge rod – A piece of straight timber marked out with saw cuts for each brick course.

Glasses – PPE that look like normal glasses and fit over the ears and nose to stay in place.

Goggles – PPE that covers the whole of the eyes with a flexible plastic shield.

Hacksaw – A frame with a handle that holds a toothed blade used to cut metals.

Half-brick wall – This is a wall which has a depth or thickness of only half a brick.

Hand tools – Tools used in the hands to carry out a task.

HASAWA – Health and Safety at Work Act 1974.

Hazards – A danger, such as a loose hammer head.

High-visibility – Uses a contrast colour, usually yellow or orange, that is visible to the eye.

Housekeeping – Cleaning up after working.

HSE – Health and Safety Executive.

Infrastructure – The structures, such as roads and railways that allow movement of goods and services.

Joint – A physical connection formed in timber.

Key – Roughening a surface so the next coat will bond to it.

Knotting – Used to seal knots in timber to prevent resin seeping out.

Lifestyle – The way in which you choose to live, work and play.

Load – Putting paint onto the brush to apply it to a surface.

Maintenance – Making sure that a tool is in good order and not broken.

Mandatory – Must be followed.

Manual handling – Using your hands and arms to lift items and not by mechanical means.

Matching pattern – Making sure that each piece of wallpaper lines up with its neighbour.

MDF – Medium Density Fibreboard.

Mitre – An angled cut in a piece of timber.

Mortar – A mixture of sand, cement and water used to rest bricks on so they are level.

Moulding – A piece of timber specially shaped to fit around something.

Multimeter – An electrical instrument that can read different voltages and amps.

Olive – A small copper ring that fits between the pipe and fitting. When tightened, a seal is formed.

Open – Behaving in a friendly manner with no barriers.

Operations – A task, for example isolating a circuit.

Pallet – The timber base that is used to lift and move items by forklift.

Paring – Using a chisel by hand to carefully remove sections of timber.

Paste – Used to fix wallpaper to a surface.

Pasting table – A table used to spread wallpaper out while you apply the paste.

Pilot hole – A small hole used to make sure that a screw is fixed in the correct place.

Plane – A tool used to shave timber to make it smooth.

Planing – Shaving timber to make it smooth.

Plug – A device used to connect equipment to the mains.

Plumb – Straight and not leaning in any way.

PPE – Personal Protective Equipment.

Preparation – The work that has to be done before you can apply any paint.

Prohibited – Not allowed.

PVA – Polyvinyl acetate glue.

Radiator – A panel which is hollow, fills with hot water and emits heat.

Render – A sand and cement mixture that acts as the first coat before plaster.

Re-pointing – Removing old mortar and filling with new.

Responsibly – Sensible and with regard for others.

Rig – A small-scale pipework assessment piece.

Sack barrows – A two-wheeled trolley with two handles.

Scale – A fraction of what the original size was.

Section – A cut through an object to show the inside.

Self-employed – Working for yourself, rather than being employed by a company.

Set square – A piece of drawing equipment that allows 90° angles to be drawn.

Shock – A discharge of electricity through your body to the earth causing muscle contraction.

Skilled – A trade skill that has taken a period of training to achieve.

SMEs – Small and medium enterprises.

Soaking – The amount of time you leave the paste to soak into the wallpaper.

Soldering – Using a blow torch and melting solder to seal joints in pipes.

SSW (safe system of work) – A system to avoid accidents at work.

Stepladders – A series of steps with side rails hinged together to form a ladder.

Stool – A small flat-topped seat with three legs.

Sub-contractors – These are contractors who work for the main builder and provide specialist services such as electrical installation works.

Substances – Chemicals, liquids and any material that could harm.

Team – Working with more than one person.

Technician – A skilled technical person who completes the main work.

Title – The name given to something, in this case a drawing.

Training – Information and instructions given in a formal way.

Veneer – A very thin sheet of timber.

Vice – A moving clamp fixed to the side of a bench for holding timber tightly.

Index